Student Solutions Manual

Michael J. Kenney
Crabtree + Company

Cynthia K. Anderson
Robert E. Lee High School

to accompany

CHEMISTRY
Matter and Its Changes
Third Edition

James E. Brady
St. John's University, New York

Joel W. Russell
Oakland University, Michigan

John R. Holum
Augsburg College (Emeritus), Minnesota

John Wiley & Sons, Inc.
New York / Chichester / Weinheim / Brisbane / Singapore / Toronto

COVER ILLUSTRATOR   Norm Christensen
COVER PHOTOS:
Welder—©Lynn Johnson/AURORA.
Fireworks—©Tony Wiles/Tony Stone Images/New York, Inc.
Hot-air balloon—©Donovan Reese/Tony Stone Images/ New York, Inc.
Lighthouse—©John Lund/Tony Stone Images/New York, Inc.
Diamond—©Color Day Productions/The Image Bank

To order books or for customer service call 1-800-CALL-WILEY (225-5945).

ISBN 0-471-35858-4

Printed in the United States of America

10 9 8 7 6 5 4 3 2 1

Printed and bound by Bradford & Bigelow, Inc.

Table of Contents

Foreword

As a student you have no doubt heard your instructor tell you the importance of practicing if you want to master the art of problem solving, a skill used frequently by chemists and all scientists. By purchasing this solution manual you affirm that statement and make a commitment to yourself that you will practice so that you will better develop your problem solving skills.

In attempting to solve any problem, there are certain steps to follow that will help you to reach a solution to the problem. First, read the entire problem; don't jump right into determining the solution. While in some cases you will immediately know the answer, it is more likely that you will need to think about the process you will need to find the solution before you will have the answer. Second, reread the problem recording any data that is presented in the problem. Then, determine, in as few words as possible, what is the question being asked. The question has been worded in a manner that makes complete sense to the questioner but may be confusing to you. Simplify the problem to a question that you understand.

At this point, it is helpful to use the method called dimensional analysis or factor–label method. This problem solving technique requires you to use both numeric values and units (dimensions) for every data presented in the problem. The challenge in identifying a solution becomes a puzzle where the solver attempts to arrange the data in a fashion that provides a solution possessing the correct units. Mastering this technique will help you throughout your studies in Chemistry.

Of course, it is very helpful and, often, imperative that you understand how to use the equations presented in the text. The early part of each chapter is arranged in a manner that will enable you to know which equation is needed to solve a specific problem. Later problems will require you to determine which equation is needed. As you go deeper into the book, you will see that it may be necessary to combine multiple equations and even to use information presented earlier in the text. Remember that Chemistry is an integrated science. Much of the material you will study depends upon basic material presented early in your studies. That material serves as the foundation upon which your chemical knowledge will be built.

How can you use this manual, then, to your advantage? In every problem we have attempted to use a consistent method for solving the problem. We try to use the factor–label method whenever appropriate or to use a specific equation if that is the correct course of action. By modeling the method we use, you will develop the skills needed to solve a variety of problems. It is important to note, though, that you may approach the problem differently than we did. That does not mean that your solution is incorrect. Quite the opposite is true. It simply indicates that you approached the problem in a different manner. Ultimately, did our solutions agree?

Be forewarned however, it is easy to misuse this manual. When attempting to solve a problem, it is extremely important that you make every effort to come to a solution without resorting to reading the solution we provide. Only after you have made significant attempts at solving the problem should you read our solution. Use it as a guide as you attempt to solve another similar problem.

In closing, we wish you well as you pursue your study of Chemistry. We hope this manual will help you. We have made every effort to provide solutions that are complete and accurate. Nonetheless, we know that errors will be present. We hope that you will let us know about these errors so that we can make corrections for future editions.

Again, good luck to you. Remember that complete understanding of Chemistry rarely comes easily. It is only through continual practice that you will master the subject.

Michael J. Kenney and Cynthia K. Anderson, November 1999

Practice Exercises

P1.1 (a)   mg   (b) μm   (c)   ps
      (a) 1×10^{-9} (b) 1×10^{-2} (c) 1×10^{-12}
      (a)   cg   (b)   Mm   (c)   ns

P1.2 $t_C = \left(t_F - 32\ ^\circ F\right)\left(\dfrac{5\ ^\circ C}{9\ ^\circ F}\right) = \left(50\ ^\circ F - 32\ ^\circ F\right)\left(\dfrac{5\ ^\circ C}{9\ ^\circ F}\right) = 10\ ^\circ C$

To convert from °F to K we first convert to °C.

$t_C = \left(t_F - 32\ ^\circ F\right)\left(\dfrac{5\ ^\circ C}{9\ ^\circ F}\right) = \left(68\ ^\circ F - 32\ ^\circ F\right)\left(\dfrac{5\ ^\circ C}{9\ ^\circ F}\right) = 20\ ^\circ C$

$T_K = 273 + t_C = 273 + 20 = 293\ K$

P1.3 (a) $\#\ in. = \left(3.00\ yd\right)\left(\dfrac{3\ ft}{1\ yd}\right)\left(\dfrac{12\ in.}{1\ ft}\right) = 108\ in.$

(b) $\#\ cm = \left(1.25\ km\right)\left(\dfrac{1000\ m}{1\ km}\right)\left(\dfrac{100\ cm}{1\ m}\right) = 1.25 \times 10^5\ cm$

(c) $\#\ ft = \left(3.27\ mm\right)\left(\dfrac{1\ m}{1000\ mm}\right)\left(\dfrac{100\ cm}{1\ m}\right)\left(\dfrac{1\ in.}{2.54\ cm}\right)\left(\dfrac{1\ ft}{12\ in.}\right) = 0.0107\ ft$

(d) $\dfrac{\#\ km}{L} = \left(\dfrac{20.2\ mile}{1\ gal}\right)\left(\dfrac{1.609\ km}{1\ mile}\right)\left(\dfrac{1\ gal}{3.786\ L}\right) = 8.58\ ^{km}\!/\!_L$

P1.4 $density = \dfrac{mass}{volume} = \dfrac{3.92\ g}{1.45\ mL} = 2.70\ ^g\!/\!_{mL}$

P1.5 (a) $\#\ cm^3 = \left(2.86\ g\right)\left(\dfrac{1\ cm^3}{10.5\ g}\right) = 0.272\ cm^3$

(b) $\#\ g = \left(16.3\ cm^3\right)\left(\dfrac{10.5\ g}{1\ cm^3}\right) = 171\ g$

P1.6 $sp.\ gr. = \dfrac{d_{aluminum}}{d_{water}} = \dfrac{2.70\ ^g\!/\!_{mL}}{1.00\ ^g\!/\!_{mL}} = 2.70$

$d_{aluminum} = sp.\ gr. \times d_{water} = 2.70 \times 62.4\ ^{lb}\!/\!_{ft^3} = 168\ ^{lb}\!/\!_{ft^3}$

Note: The density of water is taken from Example 1.7.

P1.7 $d_{ethyl\ acetate} = sp.\ gr. \times d_{water} = 0.902 \times 1.00\ ^g\!/\!_{mL} = 0.902\ ^g\!/\!_{mL}$

$d_{ethyl\ acetate} = sp.\ gr. \times d_{water} = 0.902 \times 8.34\ ^{lb}\!/\!_{gal} = 7.52\ ^{lb}\!/\!_{gal}$

Note: The density of water is taken from Example 1.7.

Review Problems

1.43 (a) 0.01 m
      (b) 1000 m
      (c) 1×10^{12} pm
      (d) 0.1 m
      (e) 0.001 kg
      (f) 0.01 g

1.45 (a) °F = 9/5 (°C) + 32= 9/5(60) + 32 = 140 °F when rounded to the proper number of significant figures.
      (b) °F = 9/5 (°C) + 32= 9/5(20) + 32 =68 °F
      (c) °C = 5/9 (°F – 32)= 5/9(45.5 –32) = 7.5 °C
      (d) °C = 5/9 (°F – 32)= 5/9(59 – 32) = 15 °C
      (e) K = °C + 273 = 40 + 273 = 313K
      (f) K = °C + 273 = –20 + 273 = 253K

1.47 °F = 9/5 (°C) + 32= 9/5(37.46) +32 = 99.43 °F.
      Since normal body temperature is 98.6 °F, this person has a fever.

1.49 °C = K – 273= 4 – 273 = –269 °C
      °F = 9/5 (°C) + 32= 9/5(–269) +32 = –452 °F

1.51 K= °C + 273 = –196 + 273 = 77K

1.53 (a) four   (b) five   (c) four
      (d) two   (e) four   (f) one

1.55 (a) 0.69   (b) 83.24   (c) 0.006
      (d) 22.84   (e) 775.4 (Note: the subtraction gives 107.3, so 775.1 is also a possible answer)

1.57 (a) 2.34×10^3 (b) 3.10×10^7 (c) 2.87×10^{-4}
      (d) 4.50×10^4 (e) 4.00×10^{-6} (f) 3.24×10^5

1.59 (a) 210,000   (b) 0.00000335   (c) 3800
      (d) 0.0000000000046   (e) 0.00000346   (f) 850,000,000

1.61 (a) 2.0×10^4
      (b) 8.0×10^7
      (c) 1.0×10^3
      (d) 2.4×10^5
      (e) 2.0×10^{18}

1.63 (a) $\# \, km = (32.0 \, dm)\left(\dfrac{1\,m}{10\,dm}\right)\left(\dfrac{1\,km}{1000\,m}\right) = 3.20 \times 10^{-3} \, km$

      (b) $\# \, \mu g = (8.2 \, mg)\left(\dfrac{1\,g}{1000\,mg}\right)\left(\dfrac{1 \times 10^6 \, \mu g}{1\,g}\right) = 8.2 \times 10^3 \, \mu g$

      (c) $\# \, kg = (75.3 \, mg)\left(\dfrac{1\,g}{1000\,mg}\right)\left(\dfrac{1\,kg}{1000\,g}\right) = 7.53 \times 10^{-5} \, kg$

      (d) $\# \, L = (137.5 \, mL)\left(\dfrac{1\,L}{1000\,mL}\right) = 0.1375 \, L$

      (e) $\# \, mL = (0.025 \, L)\left(\dfrac{1000\,mL}{1\,L}\right) = 25 \, mL$

(f) $\# \, dm = (342 \, pm)\left(\dfrac{1 \times 10^{-12} \, m}{1 \, pm}\right)\left(\dfrac{10 \, dm}{1 \, m}\right) = 3.42 \times 10^{-9} \, dm$

1.65   (a) $\# \, cm = (36 \, in.)\left(\dfrac{2.54 \, cm}{1 \, in.}\right) = 91 \, cm$

       (b) $\# \, kg = (5.0 \, lb)\left(\dfrac{1 \, kg}{2.205 \, lb}\right) = 2.3 \, kg$

       (c) $\# \, mL = (3.0 \, qt)\left(\dfrac{946.4 \, mL}{1 \, qt}\right) = 2800 \, mL$

       (d) $\# \, mL = (8 \, oz)\left(\dfrac{29.6 \, mL}{1 \, oz}\right) = 200 \, mL$

       (e) $\# \, km/hr = (55 \, mi/hr)\left(\dfrac{1.609 \, km}{1 \, mi}\right) = 88 \, km/hr$

       (f) $\# \, km = (50.0 \, mi)\left(\dfrac{1.609 \, km}{1 \, mi}\right) = 80.4 \, km$

1.67 $\# \, mL = (12 \, oz)\left(\dfrac{29.6 \, mL}{1 \, oz}\right) = 360 \, mL$

1.69 $\# \, lb = (1000 \, kg)\left(\dfrac{2.205 \, lb}{1 \, kg}\right) = 2205 \, lb$

1.71   First, 6 ft. 2 in. = 74 in.

       $\# \, cm = (74 \, in.)\left(\dfrac{2.54 \, cm}{1 \, in.}\right) = 190 \, cm$

1.73   a) $\# \, m^2 = (6.2 \, yd^2)\left(\dfrac{36 \, in}{1 \, yd}\right)^2\left(\dfrac{2.54 \, cm}{1 \, in}\right)^2\left(\dfrac{1 \, m}{100 \, cm}\right)^2 = 5.2 \, m^2$

             $\# \, mm^2 = (4.8 \, in.^2)\left(\dfrac{2.54 \, cm}{1 \, in}\right)^2\left(\dfrac{10 \, mm}{1 \, cm}\right)^2 = 3.1 \times 10^3 \, mm^2$

             $\# \, L = (3.7 \, ft^3)\left(\dfrac{12 \, in}{1 \, ft}\right)^3\left(\dfrac{2.54 \, cm}{1 \, in}\right)^3\left(\dfrac{1 \, L}{1000 \, cm^3}\right) = 1.0 \times 10^2 \, L$

1.75 $\# \, \dfrac{km}{hr} = \left(\dfrac{2235 \, ft}{1 \, s}\right)\left(\dfrac{3600 \, s}{1 \, hr}\right)\left(\dfrac{12 \, in.}{1 \, ft}\right)\left(\dfrac{2.54 \, cm}{1 \, in.}\right)\left(\dfrac{1 \, m}{100 \, cm}\right)\left(\dfrac{1 \, km}{1000 \, m}\right) = 2452 \, \dfrac{km}{hr}$

1.77   density = mass/ volume = 36.4 g/45.6 mL = 0.798 g/mL

1.79   #mL = (25.0g) /(0.791g/mL) = 31.6 mL

1.81   #mass = (1.492 g/mL)(185 mL) = 276 g

1.83   mass of silver = 62.00g – 27.35g = 34.65 g
        volume of silver = 18.3 mL –15 mL = 3.3 mL
        density of silver = (mass of siver)/(volume of silver) = (34.65 g)/(3.3 mL) = 11 g/mL

1.85

$$\text{sp.gr.} = \left(\frac{d_{substance}}{d_{water}}\right) = \left(\frac{0.715\,\text{g}/_{mL}}{1.00\,\text{g}/_{mL}}\right) = 0.715$$

1.87   Use equation 1.10 to solve this problem:

$d_{substance} = (\text{sp. gr.})_{substance} \times d_{water} = 1.47 \times 0.998\text{g/mL} = 1.47\text{ g/mL}$

mass = 1000 mL x 1.47 g/mL = 1470 g

1.89   The density of gold is 1.20×10^3 lb/ft^3

lbs = $(1\text{ft}^3)(1.20 \times 10^3\text{ lb/ft}^3) = 1.20 \times 10^3$ lb

Additional Exercises

1.93   (a)   In order to determine the volume of the pycnometer, we need to determine the volume of the water that fills it. We will do this using the mass of the water and its density.

mass of water = mass of filled pycnometer – mass of empty pycnometer

= 41.428g – 27.314g = 9.52 8g

$$\text{volume} = (9.528\text{ g})\left(\frac{1\,\text{mL}}{0.99704\,\text{g}}\right) = 9.556\,\text{mL}$$

       (b)   We know the volume of chloroform from part (a). The mass of chloroform is determined in the same way that we determined the mass of water.

Mass of chloroform = mass of filled pycnometer – mass of empty pycnometer = 41.428g – 27.314g = 14.114g

$$\text{Density of chloroform} = \left(\frac{14.114\text{g}}{9.56\text{mL}}\right) = 1.477\text{g/mL}$$

1.95 $\#\text{francs} = \#\text{s} = (186\text{ francs cabbage})\left(\frac{1\,\text{cabbage}}{42\,\text{francs}}\right)\left(\frac{3\,\text{cans potatoes}}{1\,\text{cabbage}}\right)\left(\frac{26\,\text{francs}}{1\,\text{can potatoes}}\right) = 345\text{ francs}$

1.97 $d_{sea\,water} = (1.025)\left(\frac{62.4\,\text{lb}}{\text{ft}^3}\right) = 64.0\,{\text{lb}}/_{\text{ft}^3}$

$\#\text{ ft}^3 = (4255\text{ tons})\left(\frac{2000\,\text{lbs}}{1\,\text{ton}}\right)\left(\frac{1\,\text{ft}^3}{64.0\,\text{lb}}\right) = 1.331 \times 10^5\text{ ft}^3$

1.101   Since the density closely matches the known value, we conclude that this is an authentic sample of ethylene glycol.

1.104   First we need to recall the formula which describes the circumference of a circle:

Circumference = πd = π(12 in) = 37.7 in. The circumference is the total distance a point on the edge travels in each revolution. To calculate speed:

$$\frac{\#\text{ miles}}{\text{hr}} = \left(\frac{33\,\frac{1}{3}\,\text{revolutions}}{1\,\text{minute}}\right)\left(\frac{37.7\,\text{in.}}{1\,\text{revolution}}\right)\left(\frac{2.54\,\text{cm}}{1\,\text{in.}}\right)\left(\frac{1\,\text{m}}{100\,\text{cm}}\right)\left(\frac{1\,\text{km}}{1000\,\text{m}}\right)\left(\frac{1\,\text{mile}}{1.609\,\text{km}}\right)\left(\frac{60\,\text{min}}{1\,\text{hr}}\right) = 1.2\,{\text{miles}}/_{\text{hr}}$$

Practice Exercises

P2.1   (a)   1 Ni, 2 Cl
          (b)   1 Fe, 1 S, 4 O
          (c)   3 Ca, 2 P, 8 O
          (d)   1 Co, 2 N, 12 O, 12 H

P2.2   This is a balanced chemical equation, and the number of each atom that appears on the left is the same as that on the right: 1 Mg, 2 O, 4 H, and 2 Cl.

P2.3 $Mg(OH)_2(s) + 2 HCl(aq) \rightarrow MgCl_2(aq) + 2H_2O(\ell)$

P2.4 2.24845×12 u $= 26.9814$ u

P2.5   Copper is 63.546 u $\div 12$ u $= 5.2955$ times as heavy as carbon.

P2.6 $(0.198 \ 10.0129$ u$) + (0.802 \ 11.0093$ u$) = 10.8$ u

P2.7   The atomic number (94) is equal to the number of protons, and the mass number (94 + 146 = 240) is equal to the number of protons and neutrons:
$$^{240}_{94}Pu$$

P2.8   This has 17 protons, 17 electrons, and 35 − 17 = 18 neutrons.

P2.9   (a)   K, Ar, Al (b)   Cl   (c)   Ba
          (d)   Ne   (e)   Li   (f)   Ce

P2.10   (a)   NaF   (b) Na_2O (c) MgF_2 (d) Al_4C_3

P2.11   (a) $CrCl_3$ and $CrCl_2$, Cr_2O_3 and CrO
          (b) $CuCl$, $CuCl_2$, Cu_2O and CuO

P2.12   (a) Na_2CO_3 (b) $(NH_4)_2SO_4$ (c) $KC_2H_3O_2$
          (d) $Sr(NO_3)_2$ (e) $Fe(C_2H_3O_2)_3$

P2.13   (a)   potassium sulfide   (b)   magnesium phosphide
          (c)   nickel(II) chloride   (d)   iron(III) oxide

P2.14   (a) Al_2S_3 (b) SrF_2 (c) TiO_2 (d) $CrBr_2$

P2.15   phosphorus trichloride, sulfur dioxide, dichlorine heptaoxide

P2.16   The acids are hydrofluoric acid and hydrobromic acid. The salts are sodium fluoride and sodium bromide.

P2.17   sodium arsenate

P2.18 $NaHSO_3$, sodium hydrogen sulfite

P2.19   (a)   zinc(II) nitrate
          (b)   arsenic trifluoride

Review Problems

2.117   An authentic sample of laughing gas must have a mass ratio of nitrogen/oxygen of 1.75 to 1.00. The only possibility in this list is item (c), which has the ratio of mass of nitrogen to mass of oxygen of 8.84/5.05 = 1.75.

2.119   From the first ratio we see that there is a ratio of 4.67 to 1.00. Multiplying the mass of hydrogen by 4.67 we see that for every 6.28 g hydrogen there will be 29.3 g nitrogen.

2.121   From the ratio above we see that for every 4.56 g nitrogen there need be 0.98 g hydrogen. According to the Law of Conservation of Mass then there will be 5.54 g ammonia produced.

2.123   The amount of oxygen per gram of nitrogen in NO_2 should be exactly twice that of NO, as required by the formulas of the two substances. Therefore, 2.285 g oxygen would combine with 1.000 g nitrogen.

2.125 $(12)1.6605402 \times 10^{-24}$ g $= 1.9926482 \times 10^{-23}$ g C

2.127   Since we know that the formula is CH4, we know that one fourth of the total mass due to the hydrogen atom constitutes the mass that may be compared to the carbon. Hence we have 0.33597 g H ÷ 4 = 0.083993 g H and 1.00 g assigned to the amount of C-12 in the compound. Then it is necessary to realize that the ratio 1.00 g C ÷ 12 for carbon is equal to the ratio 0.083993 g H ÷ X, where X equals the relative atomic mass of hydrogen.

$$\left(\frac{1.000 \text{ g C}}{12 \text{ u C}}\right) = \left(\frac{0.083993 \text{ g H}}{X}\right) = 1.008 \text{ u}$$

2.129   Regardless of the definition, the ratio of the mass of hydrogen to that of oxygen would be the same. If C–12 were assigned a mass of 24 (twice its accepted value), then hydrogen would also have a mass twice its current value, or 2.0158 u.

2.131 $(0.6916 \times 62.9396 \text{ u}) + (0.3083 \times 64.9278 \text{ u}) = 63.55 \text{ u}$

2.133

	neutrons	protons	electrons
(a)	138	88	88
(b)	8	6	6
(c)	124	82	82
(d)	12	11	11

2.135

	neutrons	protons	electrons
(a)	46	35	36
(b)	32	26	23
(c)	34	29	27
(d)	50	37	36

Additional Exercises

2.138   First calculate the number of moles of oxygen that are combined with X:

$$\left(1.14 \text{ g oxygen}\right)\left(\frac{1 \text{ mole O}}{16.00 \text{ g O}}\right) = 0.0713 \text{ moles oxygen}$$

Then calculate the moles of X in the oxygen compound:

$$\left(0.0713 \text{ moles O}\right)\left(\frac{1 \text{ mole X}}{2 \text{ moles O}}\right) = 0.0356 \text{ moles X}$$

Finally, calculate the molar mass of X:

$$\frac{1.00\,g\,X}{0.0356\,moles\,X} = 28.1\,g/mole\,X$$

Note that X is Si, atomic number 14, and that its oxygen compound is SiO_2.

The same number of moles of X are also combined with Y:

$$(0.0356\,moles\,X)\left(\frac{4\,moles\,Y}{1\,moles\,X}\right) = 0.142\,moles\,Y$$

The atomic mass of Y is thus:

$$\frac{5.07\,g\,Y}{0.142\,moles\,Y} = 35.6\,g/mole\,Y$$

Note that Y is Cl, atomic number 17, and that its compound with X is $SiCl_4$.

2.142 Na_2HPO_4 and NaH_2PO_4

2.146

$$\# gal\,O_2 = (385,200\,gal\,H_2)\left(\frac{3.785\,L}{1\,gal}\right)\left(\frac{1000\,mL}{1\,L}\right)\left(\frac{0.0708\,g\,H_2}{1\,mL}\right)\left(\frac{8\,g\,O_2}{1\,g\,H_2}\right)$$

$$\times \left(\frac{1\,mL}{1.139\,g\,O_2}\right)\left(\frac{1\,L}{1000\,mL}\right)\left(\frac{1\,gal}{3.785\,L}\right) = 191,600\,gal$$

Practice Exercises

P3.1 $\# \text{mol N} = (8.60 \text{ mol O})\left(\dfrac{2 \text{ mol N}}{5 \text{ mol O}}\right) = 3.44 \text{ mol N}$

P3.2 $\# \text{mol S} = (35.6 \text{ g S})\left(\dfrac{1 \text{ mol S}}{32.07 \text{ g S}}\right) = 1.11 \text{ mol S}$

P3.3 $\# \text{g Ag} = (0.263 \text{ mol Ag})\left(\dfrac{107.9 \text{ g Ag}}{1 \text{ mol Ag}}\right) = 28.4 \text{ g Ag}$

P3.4 $\# \text{atoms Pb} = (1.000 \times 10^{-9} \text{ g Pb})\left(\dfrac{1 \text{ mol Pb}}{207.2 \text{ g Pb}}\right)\left(\dfrac{6.022 \times 10^{23} \text{ atoms Pb}}{1 \text{ mol Pb}}\right)$

$= 2.906 \times 10^{12} \text{ atoms Pb}$

P3.5 $Na_2CO_3 = 2 Na + C + 3 O$

Formula mass of $Na_2CO_3 = (2 \times 22.99) + (12.01) + (3 \times 16.00)$

$= 45.98 + 12.01 + 48.00$

$= 105.99 \text{ g}$

P3.6 $0.125 \text{ mol} \times 106 \text{ g/mol} = 13.3 \text{ g}$

P3.7 $\# \text{mol H}_2\text{SO}_4 = (45.8 \text{ g H}_2\text{SO}_4)\left(\dfrac{1 \text{ mol H}_2\text{SO}_4}{98.1 \text{ g H}_2\text{SO}_4}\right) = 0.467 \text{ mol H}_2\text{SO}_4$

P3.8 $\# \text{g Fe} = (25.6 \text{ g O})\left(\dfrac{1 \text{ mol O}}{16.0 \text{ g O}}\right)\left(\dfrac{2 \text{ mol Fe}}{3 \text{ mol O}}\right)\left(\dfrac{55.8 \text{ g Fe}}{1 \text{ mol Fe}}\right) = 59.5 \text{ g Fe}$

P3.9   % N = 0.1417/0.5462 × 100 = 25.94 % N
% O = 0.4045/0.5462 × 100 = 74.06 % O
Since these two values constitute 100 %, there are no other elements present.

P3.10   We first determine the number of grams of each element that are present in one mole of sample:
2 mol N × 14.01 g/mol = 28.02 g N
3 mol O × 16.00 g/mol = 48.00 g O
The percentages by mass are then obtained using the formula mass of the compound (76.02 g):
% N = 28.02/76.02 × 100 = 36.86 % N
% O = 48.00/76.02 × 100 = 63.14 % O

P3.11 $\# \text{g Fe} = (15.0 \text{ g Fe}_2\text{O}_3)\left(\dfrac{111.7 \text{ g Fe}}{159.7 \text{ g Fe}_2\text{O}_3}\right) = 10.5 \text{ g Fe}$

P3.12   We first determine the number of moles of each element as follows:

$\# \text{mol N} = (0.7117 \text{ g N})\left(\dfrac{1 \text{ mol N}}{14.01 \text{ g N}}\right) = 0.05080 \text{ mol N}$

We need to know the number of grams of O. Since there is a total of 1.5246 g of compound and the only other element present is N, the mass of O = 1.5246 g − 0.7117 g = 0.8129 g.

$$\# \text{ mol O} = \left(0.8129\,\text{g O}\right)\left(\frac{1\,\text{mol O}}{16.00\,\text{g O}}\right) = 0.05081\,\text{mol O}$$

Since these two mole amounts are the nearly identical, the empirical formula is NO.

P3.13   First convert the number of grams of each element into number of moles that are present in the compound:

$$\# \text{ mol N} = \left(0.5219\,\text{g N}\right)\left(\frac{1\,\text{mol N}}{14.01\,\text{g N}}\right) = 0.03725\,\text{mol N}$$

The number of grams of O = 2.012 g – 0.5219 g = 1.490 g.

$$\# \text{ mol O} = \left(1.490\,\text{g O}\right)\left(\frac{1\,\text{mol O}}{16.00\,\text{g O}}\right) = 0.09313\,\text{mol O}$$

We next divide these two mole amounts by the smaller of the two, in order to determine the smallest whole number ratio:

For N: 0.03725 mol/0.03725 mol = 1.000
For O: 0.09313 mol/0.03725 mol= 2.500

By inspection, this ratio is seen to be the same as 2:5, and we deduce that the empirical formula is N_2O_5.

P3.14   It is convenient to assume that we have 100 g of the sample, so that the % by mass values may be taken directly to represent masses. Thus there is 32.37 g of Na, 22.57 g of S and (100.00 – 32.37 – 22.57) = 45.06 g of O. Now, convert these masses to a number of moles:

$$\# \text{ mol Na} = \left(32.37\,\text{g Na}\right)\left(\frac{1\,\text{mol Na}}{23.00\,\text{g Na}}\right) = 1.407\,\text{mol Na}$$

$$\# \text{ mol S} = \left(22.57\,\text{g S}\right)\left(\frac{1\,\text{mol S}}{32.06\,\text{g S}}\right) = 0.7040\,\text{mol S}$$

$$\# \text{ mol O} = \left(45.06\,\text{g O}\right)\left(\frac{1\,\text{mol O}}{16.00\,\text{g O}}\right) = 2.816\,\text{mol O}$$

Next, we divide each of these mole amounts by the smallest in order to deduce the simplest whole number ratio:

For Na: 1.407 mol/0.7040 mol = 1.999
For S: 0.7040 mol/0.7040 mol = 1.000
For O: 2.816 mol/0.7040 mol = 4.000

The empirical formula is Na_2SO_4.

P3.15   Since the entire amount of carbon that was present in the original sample appears among the products only as CO_2, we calculate the amount of carbon in the sample as follows:

$$\# \text{ g C} = \left(7.406\,\text{g CO}_2\right)\left(\frac{1\,\text{mol CO}_2}{44.01\,\text{g CO}_2}\right)\left(\frac{1\,\text{mol C}}{1\,\text{mol CO}_2}\right)\left(\frac{12.01\,\text{g C}}{1\,\text{mol C}}\right) = 2.021\,\text{g C}$$

Similarly, the entire mass of hydrogen that was present in the original sample appears among the products only as H_2O. Thus the mass of hydrogen in the sample is:

$$\# \text{ g H} = \left(3.027\,\text{g H}_2\text{O}\right)\left(\frac{1\,\text{mol H}_2\text{O}}{18.02\,\text{g H}_2\text{O}}\right)\left(\frac{2\,\text{mol H}}{1\,\text{mol H}_2\text{O}}\right)\left(\frac{1.008\,\text{g H}}{1\,\text{mol H}}\right) = 0.3386\,\text{g H}$$

The mass of oxygen in the original sample is determined by difference:

$$5.048 \text{ g} - 2.021 \text{ g} - 0.3386 \text{ g} = 2.688 \text{ g O}$$

Next, these mass amounts are converted to the corresponding mole amounts:

$$\# \text{ mol C} = (2.021 \text{ g C})\left(\frac{1 \text{ mol C}}{12.01 \text{ g C}}\right) = 0.1683 \text{ mol C}$$

$$\# \text{ mol H} = (0.3386 \text{ g H})\left(\frac{1 \text{ mol H}}{1.008 \text{ g H}}\right) = 0.3359 \text{ mol H}$$

$$\# \text{ mol O} = (2.688 \text{ g O})\left(\frac{1 \text{ mol O}}{16.00 \text{ g O}}\right) = 0.1680 \text{ mol O}$$

The simplest formula is obtained by dividing each of these mole amounts by the smallest:

For C: 0.1683 mol/0.1680 mol = 1.002
for H: 0.3359 mol/0.1680 mol = 1.999
For O: 0.1680 mol/0.1680 mol = 1.000

These values give us the simplest formula directly, namely CH_2O.

P3.16 The formula mass of the empirical unit is $1 \text{ N} + 2 \text{ H} = 16.03$. Since this is half of the molecular mass, the molecular formula is N_2H_4.

P3.17 $3CaCl_2(aq) + 2K_3PO_4(aq) \rightarrow Ca_3(PO_4)_2(s) + 6KCl(aq)$

P3.18 $2Fe(s) + 3Cl_2(g) \rightarrow 2FeCl_3(s)$

P3.19 $\# \text{ mol H}_2SO_4 = (0.366 \text{ mol NaOH})\left(\frac{1 \text{ mol H}_2SO_4}{2 \text{ mol NaOH}}\right) = 0.183 \text{ mol H}_2SO_4$

P3.20 $\# \text{ mol O}_2 = (0.575 \text{ mol CO}_2)\left(\frac{5 \text{ mol O}_2}{3 \text{ mol CO}_2}\right) = 0.958 \text{ mol O}_2$

P3.21 $\# \text{ g Al}_2O_3 = (86.0 \text{ g Fe})\left(\frac{1 \text{ mol Fe}}{55.85 \text{ g Fe}}\right)\left(\frac{1 \text{ mol Al}_2O_3}{2 \text{ mol Fe}}\right)\left(\frac{102.0 \text{ g Al}_2O_3}{1 \text{ mol Al}_2O_3 \text{ mol Al}_2O_3}\right)$

$= 78.5 \text{ g Al}_2O_3$

P3.22 First determine the number of grams of O_2 that would be required to react completely with the given amount of ammonia:

$$\# \text{ g O}_2 = (30.00 \text{ g NH}_3)\left(\frac{1 \text{ mol NH}_3}{17.03 \text{ g NH}_3}\right)\left(\frac{5 \text{ mol O}_2}{4 \text{ mol NH}_3}\right)\left(\frac{32.00 \text{ g O}_2}{1 \text{ mol O}_2}\right)$$

$$= 70.46 \text{ g O}_2$$

Since this is more than the amount that is available, we conclude that oxygen is the limiting reactant. The rest of the calculation is therefore based on the available amount of oxygen:

$$\# \text{ g NO} = (40.00 \text{ g O}_2)\left(\frac{1 \text{ mol O}_2}{32.00 \text{ g O}_2}\right)\left(\frac{4 \text{ mol NO}}{5 \text{ mol O}_2}\right)\left(\frac{30.01 \text{ g NO}}{1 \text{ mol NO}}\right)$$

$$= 30.01 \text{ g NO}$$

P3.23   First determine which reactant is limiting by calculating the number of grams of $Na_2Cr_2O_7$ that are required to react with the given amount of alcohol:

$$\# g\, Na_2Cr_2O_7 = (24.0\,g\, C_2H_6O)\left(\frac{1\,mol\, C_2H_6O}{46.07\,g\, C_2H_6O}\right)$$

$$\times \left(\frac{2\,mol\, Na_2Cr_2O_7}{3\,mol\, C_2H_6O}\right)\left(\frac{261.99\,g\, Na_2Cr_2O_7}{1\,mol\, Na_2Cr_2O_7}\right)$$

$$= 91.0\,g\, Na_2Cr_2O_7$$

Since this is slightly more than the amount of $Na_2Cr_2O_7$ that has been made available, we conclude that $Na_2Cr_2O_7$ is the limiting reactant. We then use the amount of $Na_2Cr_2O_7$ present in order to calculate a theoretical yield.

$$\# g\, C_2H_4O_2 = (90.0\,g\, Na_2Cr_2O_7)\left(\frac{1\,mol\, Na_2Cr_2O_7}{261.99\,g\, Na_2Cr_2O_7}\right)$$

$$\times \left(\frac{3\,mol\, C_2H_4O_2}{2\,mol\, Na_2Cr_2O_7}\right)\left(\frac{60.05\,g\, C_2H_4O_2}{1\,mol\, C_2H_4O_2}\right)$$

$$= 30.9\,g\, C_2H_4O_2$$

The percent yield is then obtained by dividing the actual yield by the theoretical yield:
26.6 g/30.9 g × 100 = 86.1 %

P3.24 $$\frac{\#\,moles}{L} = \left(\frac{3.550\,g\, Na_2SO_4}{100.0\,mL}\right)\left(\frac{1\,mol\, Na_2SO_4}{142.06\,g\, Na_2SO_4}\right)\left(\frac{1000\,mL}{1\,L}\right) = 0.2499\,M$$

P3.25 $$\#\,mL\,solution = (0.0500\,mol\, KCl)\left(\frac{1\,L\,solution}{0.150\,mol\, KCl}\right)\left(\frac{1000\,mL}{1\,L}\right)$$

$$= 333\,mL\,solution$$

P3.26

$$\# g\, CaCl_2 = (250\,mL\,solution)\left(\frac{1\,L}{1000\,mL}\right)\left(\frac{0.125\,mol\, CaCl_2}{1\,L\,solution}\right)$$

$$\times \left(\frac{110.98\,g\, CaCl_2}{1\,mol\, CaCl_2}\right)$$

$$= 3.47\,g\, CaCl_2$$

Note: The volume is assumed to be exact.

P3.27   Use equation 3.4

$$V_{concd} = \frac{V_{dil}\cdot M_{dil}}{M_{concd}} = \frac{(100\,mL)(0.125\,M\, H_2SO_4)}{0.500\,M\, H_2SO_4} = 25.0\,mL\, H_2SO_4$$

Dilute 25.0 mL of concentrated acid to a final volume of 100 mL.

P3.28

$$\text{\# mL NaOH soln} = (15.4 \text{ ml } H_2SO_4 \text{ soln})\left(\frac{0.108 \text{ mol } H_2SO_4}{1000 \text{ mL } H_2SO_4 \text{ soln}}\right)$$

$$\times \left(\frac{2 \text{ mol NaOH}}{1 \text{ mol } H_2SO_4}\right)\left(\frac{1000 \text{ mL NaOH soln}}{0.124 \text{ mol NaOH}}\right)$$

$$= 26.8 \text{ mL NaOH soln}$$

Review Problems

3.17   1:2

3.19 2.59×10^{-3} mole Na

3.21   (a)   6:11
        (b)   12:11
        (c)   2:1
        (d)   2:1

3.23 $\text{\# mol Al} = (1.58 \text{ mol O})\left(\frac{2 \text{ mol Al}}{3 \text{ mol O}}\right) = 1.05 \text{ mol Al}$

3.25 $\text{\# mol Al} = (2.16 \text{ mol } Al_2O_3)\left(\frac{2 \text{ mol Al}}{1 \text{ mol } Al_2O_3}\right) = 4.32 \text{ mol Al}$

3.27   (a) $\left(\frac{2 \text{ mol Al}}{3 \text{ mol S}}\right) \text{ or } \left(\frac{3 \text{ mol S}}{2 \text{ mol Al}}\right)$

        (b) $\left(\frac{3 \text{ mol S}}{1 \text{ mol } Al_2(SO_4)_3}\right) \text{ or } \left(\frac{1 \text{ mol } Al_2(SO_4)_3}{3 \text{ mol S}}\right)$

        (c) $\text{\# mol Al} = (0.900 \text{ mol S})\left(\frac{2 \text{ mol Al}}{3 \text{ mol S}}\right) = 0.600 \text{ mol Al}$

        (d) $\text{\# mol S} = (1.16 \text{ mol } Al_2(SO_4)_3)\left(\frac{3 \text{ mol S}}{1 \text{ mol } Al_2(SO_4)_3}\right) = 3.48 \text{ mol S}$

3.29   Based on the balanced equation:
        $$2NH_{3(g)} \rightarrow N_{2(g)} + 3H_{2(g)}$$
        From this equation the conversion factors can be written:

        $$\left(\frac{1 \text{ mol } N_2}{2 \text{ mol } NH_3}\right) \text{ and} \left(\frac{3 \text{ mol } H_2}{2 \text{ mol } NH_3}\right)$$

        To determine the moles produced, simply convert from starting moles to end moles:

        $$0.145 \text{ mol } NH_3\left(\frac{1 \text{ mol } N_2}{2 \text{ mol } NH_3}\right) = 0.0725 \text{ mol } N_2$$

        The moles of hydrogen are calculated similarly:

        $$0.145 \text{ mol } NH_3\left(\frac{3 \text{ mol } H_2}{2 \text{ mol } NH_3}\right) = 0.218 \text{ mol } H_2$$

3.31 $\# \text{ moles UF}_6 = (1.25 \text{ mol CF}_4)\left(\dfrac{4 \text{ mol F}}{1 \text{ mol CF}_4}\right)\left(\dfrac{1 \text{ mol UF}_6}{6 \text{ mol F}}\right) = 0.833 \text{ moles UF}_6$

3.33

$\# \text{ atoms C} = (3.13 \text{ mol H})\left(\dfrac{1 \text{ mol C}_3\text{H}_8}{8 \text{ mol H}}\right)\left(\dfrac{6.022 \times 10^{23} \text{ molecules C}_3\text{H}_8}{1 \text{ mol C}_3\text{H}_8}\right)$

$\left(\dfrac{3 \text{ atoms C}}{1 \text{ molecule C}_3\text{H}_8}\right) = 7.07 \times 10^{23} \text{ atoms C}$

3.35

$\# \text{ atoms} = (0.260 \text{ mol glucose})\left(\dfrac{6.022 \times 10^{23} \text{ molecules glucose}}{1 \text{ mol glucose}}\right)$

$\left(\dfrac{24 \text{ atoms}}{1 \text{ molecule glucose}}\right) = 3.76 \times 10^{24} \text{ atoms}$

3.37 $6 \text{ g} \div 12 \text{ g/mol} = 0.5 \text{ mol, or } 3.01 \times 10^{23} \text{ atoms}$

3.39   (a)   23.0 g
      (b)   32.1 g
      (c)   35.5 g

3.41   (a) $\# \text{ g Fe} = (1.35 \text{ mol Fe})\left(\dfrac{55.85 \text{ g Fe}}{1 \text{ mole Fe}}\right) = 75.4 \text{ g Fe}$

      (b) $\# \text{ g O} = (24.5 \text{ mol O})\left(\dfrac{16.0 \text{ g O}}{1 \text{ mole O}}\right) = 392 \text{ g O}$

      (c) $\# \text{ g Ca} = (0.876 \text{ mol Ca})\left(\dfrac{40.08 \text{ g Ca}}{1 \text{ mole Ca}}\right) = 35.1 \text{ g Ca}$

3.43 $\# \text{ moles Ni} = (18.7 \text{ g Ni})\left(\dfrac{1 \text{ mol Ni}}{58.69 \text{ g Ni}}\right) = 0.319 \text{ moles Ni}$

3.45   Note: all masses are in g/mole
      (a) NaHCO_3 =   1Na + 1H + 1C + 3O
                         =   (22.99) + (1.01) + (12.01) + (3 × 16.00)
                         =   84.0 g/mole
      (b) $\text{K}_2\text{Cr}_2\text{O}_7$ =   2K + 2Cr + 7O
                         =   (2 × 39.10) + (2 × 52.00) + (7 × 16.00)
                         =   294.2 g/mole
      (c) $(\text{NH}_4)_2\text{CO}_3$ =   2N + 8H + C + 3O
                         =   (2 × 14.01) + (8 × 1.01) + (12.01) + (3 × 16.00)
                         =   96.1 g/mole
      (d) $\text{Al}_2(\text{SO}_4)_3$ =   2Al + 3S + 12O
                         =   (2 × 26.98) + (3 × 32.07) + (12 × 16.00)
                         =   342.2 g/mole
      (e) $\text{CuSO}_4\cdot 5\text{H}_2\text{O}$ =   1Cu + 1S + 9O + 10H
                         =   63.55 + 32.07 + (9 × 16.00) + (10 × 1.01)
                         =   249.7 g/mole

3.47   (a)   # g =(1.25 mol $Ca_3(PO_4)_2$)(310.18 g $Ca_3(PO_4)_2$/1 mol $Ca_3(PO_4)_2$) = 388 g $Ca_3(PO_4)_2$

          (b)   # g =(0.625 mol $Fe(NO_3)_3$)(241.87 g $Fe(NO_3)_3$/1 mol $Fe(NO_3)_3$ = 151 g $Fe(NO_3)_3$

          (c)   # g = (0.600 mol C_4H_{10})(58.12 g C_4H_{10}/1 mole C_4H_{10}) = 34.9 g C_4H_{10}

          (d)   # g = (1.45 mol $(NH_4)_2CO_3$)(96.11 g/mole) = 139 g $(NH_4)_2CO_3$

3.49   (a)   # moles $CaCO_3$ = (21.5 g $CaCO_3$)$\left(\dfrac{1\,mole\,CaCO_3}{100.09\,g\,CaCO_3}\right)$ = 0.215 moles $CaCO_3$

          (b)   # moles NH_3 = (1.56 g NH_3)$\left(\dfrac{1\,mole\,NH_3}{17.03\,g\,NH_3}\right)$ = 9.16×10^{-2} moles NH_3

          (c)   # moles $Sr(NO_3)_2$ = (16.8 g $Sr(NO_3)_2$)$\left(\dfrac{1\,mole\,Sr(NO_3)_2}{211.6\,g\,Sr(NO_3)_2}\right)$

                             = 7.94×10^{-2} moles $Sr(NO_3)_2$

          (d)   # moles Na_2CrO_4 = (6.98×10^{-6} g Na_2CrO_4)$\left(\dfrac{1\,mole\,Na_2CrO_4}{162.0\,g\,Na_2CrO_4}\right)$

                             = 4.31×10^{-8} moles Na_2CrO_4

3.51   The formula CaC_2 indicates that there is 1 mole of Ca for every 2 moles of C. Therefore, if there are 0.150 moles of C there must be 0.0750 moles of Ca.

         # g Ca = (0.075 mol Ca)$\left(\dfrac{40.078\,g\,Ca}{1\,mole\,Ca}\right)$ = 3.01 g Ca

3.53   # mol N = (0.650 mol $(NH_4)_2CO_3$)$\left(\dfrac{2\,moles\,N}{1\,mole\,(NH_4)_2CO_3}\right)$ = 1.30 moles N

         # g $(NH_4)_2CO_3$ = (0.650 mol $(NH_4)_2CO_3$)$\left(\dfrac{96.09\,g\,(NH_4)_2CO_3}{1\,mole\,(NH_4)_2CO_3}\right)$

                    = 62.5 g $(NH_4)_2CO_3$

3.55   Assume one mole total for each of the following.

      (a)

          The molar mass of NaH_2PO_4 is 119.98 g/mole.

          % Na = $\dfrac{23.0\,g\,Na}{119.98\,g\,NaH_2PO_4}$ × 100 % = 19.2 %

          % H = $\dfrac{2.02\,g\,H}{119.98\,g\,NaH_2PO_4}$ × 100 % = 1.68 %

          % P = $\dfrac{31.0\,g\,P}{119.98\,g\,NaH_2PO_4}$ × 100 % = 25.8 %

          % O = $\dfrac{64.0\,g\,O}{119.98\,g\,NaH_2PO_4}$ × 100 % = 53.3 %

(b)

The molar mass of $NH_4H_2PO_4$ is 115.05 g/mole.

$$\% \, N = \frac{14.0 \, g \, N}{115.05 \, g \, NH_4H_2PO_4} \times 100 \, \% = 12.2 \, \%$$

$$\% \, H = \frac{6.05 \, g \, H}{115.05 \, g \, NH_4H_2PO_4} \times 100 \, \% = 5.26 \, \%$$

$$\% \, P = \frac{31.0 \, g \, P}{115.05 \, g \, NH_4H_2PO_4} \times 100 \, \% = 26.9 \, \%$$

$$\% \, O = \frac{64.0 \, g \, O}{115.05 \, g \, NH_4H_2PO_4} \times 100 \, \% = 55.6 \, \%$$

(c)

The molar mass of $(CH_3)_2CO$ is 58.05 g/mole.

$$\% \, C = \frac{36.0 \, g \, C}{58.05 \, g \, (CH_3)_2CO} \times 100 \, \% = 62.0 \, \%$$

$$\% \, H = \frac{6.05 \, g \, H}{58.05 \, g \, (CH_3)_2CO} \times 100 \, \% = 10.4 \, \%$$

$$\% \, O = \frac{16.0 \, g \, O}{58.05 \, g \, (CH_3)_2CO} \times 100 \, \% = 27.6 \, \%$$

(d)

The molar mass of $CaSO_4$ is 136.2 g/mole.

$$\% \, Ca = \frac{40.1 \, g \, Ca}{136.2 \, g \, CaSO_4} \times 100 \, \% = 29.4 \, \%$$

$$\% \, S = \frac{32.1 \, g \, S}{136.2 \, g \, CaSO_4} \times 100 \, \% = 23.6 \, \%$$

$$\% \, O = \frac{64.0 \, g \, O}{136.2 \, g \, CaSO_4} \times 100 \, \% = 47.0 \, \%$$

(e)

The molar mass of $CaSO_4 \cdot 2H_2O$ is 172.2 g/mole.

$$\% \, Ca = \frac{40.1 \, g \, Ca}{172.2 \, g \, CaSO_4 \cdot 2H_2O} \times 100 \, \% = 23.3 \, \%$$

$$\% \, S = \frac{32.1 \, g \, S}{172.2 \, g \, CaSO_4 \cdot 2H_2O} \times 100 \, \% = 18.6 \, \%$$

$$\% \, O = \frac{96.0 \, g \, O}{172.2 \, g \, CaSO_4 \cdot 2H_2O} \times 100 \, \% = 55.7 \, \%$$

$$\% \, H = \frac{4.03 \, g \, H}{172.2 \, g \, CaSO_4 \cdot 2H_2O} \times 100 \, \% = 2.34 \, \%$$

3.57 $$\% \, P = \frac{0.539 \, g \, P}{2.35 \, g \, compound} \times 100\% = 22.9\%$$

$\% \, Cl = 100\% - 22.9\% = 77.1\%$

3.59   For $C_{17}H_{25}N$, the molar mass (17C + 25H + 1N) equals 243.39 g/mole, and the three theoretical values for % by weight are calculated as follows:

$$\% \, C = \frac{204.2 \, g \, C}{243.4 \, g \, C_{17}H_{25}N} \times 100\% = 83.89\%$$

$$\% \, H = \frac{25.20 \, g \, H}{243.4 \, g \, C_{17}H_{25}N} \times 100\% = 10.35\%$$

$$\% \, N = \frac{14.01 \, g \, N}{243.4 \, g \, C_{17}H_{25}N} \times 100\% = 5.76\%$$

These data are consistent with the experimental values cited in the problem.

3.61 $\# \, g \, O = (7.14 \times 10^{21} \, atoms \, N) \left(\frac{1 \, mol \, N}{6.02 \times 10^{23} \, atoms \, N} \right) \left(\frac{5 \, mol \, O}{2 \, mol \, N} \right) \left(\frac{16.0 \, g \, O}{1 \, mol \, O} \right) = 0.474 \, g \, O$

3.63   The molecular formula is some integer multiple of the empirical formula. This means that we can divide the molecular formula by the largest possible whole number that gives an integer ratio among the atoms in the empirical formula.
(a)   SCl   (b) CH_2O (c) NH_3 (d) AsO_3 (e)   HO

3.65   We begin by realizing that the mass of oxygen in the compound may be determined by difference: 0.8961 g total – (0.1114 g Na + 0.4748 g Tc) = 0.3099 g O. Next we can convert each mass of an element into the corresponding number of moles of that element as follows:

$$\# \, moles \, Na = (0.111 \, g \, Na) \left(\frac{1 \, mole \, Na}{23.00 \, g \, Na} \right) = 4.83 \times 10^{-3} \, moles \, Na$$

$$\# \, moles \, Tc = (0.477 \, g \, Tc) \left(\frac{1 \, mole \, Tc}{98.9 \, g \, Tc} \right) = 4.82 \times 10^{-3} \, moles \, Tc$$

$$\# \, moles \, O = (0.308 \, g \, O) \left(\frac{1 \, mole \, O}{16.0 \, g \, O} \right) = 1.93 \times 10^{-2} \, moles \, O$$

Now we divide each of these numbers of moles by the smallest of the three numbers, in order to obtain the simplest mole ratio among the three elements in the compound:

for Na, 4.83×10^{-3} moles / 4.82×10^{-3} moles = 1.00
for Tc, 4.82×10^{-3} moles / 4.82×10^{-3} moles = 1.000
for O, 1.93×10^{-2} moles / 4.82×10^{-3} moles = 4.00

These relative mole amounts give us the empirical formula: $NaTcO_4$.

3.67   Assume a 100g sample:

$$\# \, mol \, C = (14.5 \, g \, C) \left(\frac{1 \, mol \, C}{12.01 \, g \, C} \right) = 1.21 \, mol \, C$$

$$\# \, mol \, Cl = (85.5 \, g \, Cl) \left(\frac{1 \, mol \, Cl}{35.45 \, g \, Cl} \right) = 2.41 \, mol \, Cl$$

Now we divide each of these numbers of moles by the smallest of the three numbers, in order to obtain the simplest mole ratio among the three elements in the compound:

for C, 1.21 moles / 1.21 moles = 1.00
for Cl, 2.41 moles / 1.21 moles = 2.000

These relative mole amounts give us the empirical formula CCl_2

3.69   Assume a 100g sample:

$$\# \text{ mol Na} = (22.9 \text{ g Na})\left(\frac{1 \text{ mol Na}}{22.99 \text{ g Na}}\right) = 0.996 \text{ mol Na}$$

$$\# \text{ mol B} = (21.5 \text{ g B})\left(\frac{1 \text{ mol B}}{10.81 \text{ g B}}\right) = 1.99 \text{ mol B}$$

$$\# \text{ mol O} = (55.7 \text{ g O})\left(\frac{1 \text{ mol O}}{16.00 \text{ g O}}\right) = 3.48 \text{ mol O}$$

Now we divide each of these numbers of moles by the smallest of the three numbers, in order to obtain the simplest mole ratio among the three elements in the compound:

for Na, 0.996 moles / 0.996 moles = 1.00
for B, 1.99 moles / 0.996 moles = 2.00
for O, 3.48 moles / 0.996 moles = 3.49

These relative mole amounts give us the empirical formula $Na_2B_4O_7$

3.71   All of the carbon is converted to carbon dioxide so,

$$\# \text{ g C} = (1.312 \text{ g CO}_2)\left(\frac{1 \text{ mol CO}_2}{44.01 \text{ g CO}_2}\right)\left(\frac{1 \text{ mol C}}{1 \text{ mol CO}_2}\right)\left(\frac{12.01 \text{ g C}}{1 \text{ mol C}}\right) = 0.358 \text{ g C}$$

$$\# \text{ moles C} = (0.358 \text{ g C})\left(\frac{1 \text{ mol C}}{12.01 \text{ g C}}\right) = 2.98 \times 10^{-2} \text{ mol C}$$

All of the hydrogen is converted to H_2O so,

$$\# \text{ g H} = (0.805 \text{ g H}_2\text{O})\left(\frac{1 \text{ mol H}_2\text{O}}{18.02 \text{ g H}_2\text{O}}\right)\left(\frac{2 \text{ mol H}}{1 \text{ mol H}_2\text{O}}\right)\left(\frac{1.008 \text{ g H}}{1 \text{ mol H}}\right) = 0.0901 \text{ g H}$$

$$\# \text{ mol H} = (0.0901 \text{ g H})\left(\frac{1 \text{ mol H}}{1.008 \text{ g H}}\right) = 8.93 \times 10^{-2} \text{ moles H}$$

The amount of O in the compound is determined by subtracting the mass of C and the mass of H from the sample.

g O = 0.684g – 0.358 g – 0.0901 g = 0.236 g

$$\# \text{ mol O} = (0.236 \text{ g O})\left(\frac{1 \text{ mol O}}{16 \text{ g O}}\right) = 1.48 \times 10^{-2} \text{ moles}$$

The relative mole ratios are:
for C, 0.0298 moles / 0.0148 moles = 2.01
for H, 0.0893 moles/ 0.0148 moles = 6.03
for O, 0.0148 moles / 0.0148 moles = 1.00

The relative mole amounts give the empirical formula C_2H_6O

3.73   This type of combustion analysis takes advantage of the fact that the entire amount of carbon in the original sample appears as CO_2 among the products. Hence the mass of carbon in the original sample must be equal to the mass of carbon that is found in the CO_2.

$$\# \text{ g C} = (19.73 \times 10^{-3} \text{ g CO}_2)\left(\frac{1 \text{ mole CO}_2}{44.01 \text{ g CO}_2}\right)\left(\frac{1 \text{ mole C}}{1 \text{ mole CO}_2}\right)\left(\frac{12.011 \text{ g C}}{1 \text{ mole C}}\right)$$

$$= 5.385 \times 10^{-3} \text{ g C}$$

Similarly, the entire mass of hydrogen that was present in the original sample ends up in the products as H_2O:

$$\# \text{ g H} = (6.391 \times 10^{-3} \text{ g H}_2O)\left(\frac{1 \text{ mole H}_2O}{18.02 \text{ g H}_2O}\right)\left(\frac{2 \text{ mole H}}{1 \text{ mole H}_2O}\right)\left(\frac{1.008 \text{ g H}}{1 \text{ mole H}}\right)$$

$$= 7.150 \times 10^{-4} \text{ g H}$$

The mass of oxygen is determined by subtracting the mass due to C and H from the total mass: 6.853 mg total − (5.385 mg C + 0.7150 mg H) = 0.753 mg O. Now, convert these masses to a number of moles:

$$\# \text{ moles C} = \left(5.385 \times 10^{-3} \text{ g C}\right)\left(\frac{1 \text{ mole C}}{12.011 \text{ g C}}\right) = 4.483 \times 10^{-4} \text{ moles C}$$

$$\# \text{ moles H} = \left(7.150 \times 10^{-4} \text{ g H}\right)\left(\frac{1 \text{ mole H}}{1.0079 \text{ g H}}\right) = 7.094 \times 10^{-4} \text{ moles H}$$

$$\# \text{ moles O} = \left(7.53 \times 10^{-4} \text{ g O}\right)\left(\frac{1 \text{ mole O}}{15.999 \text{ g O}}\right) = 4.71 \times 10^{-5} \text{ moles O}$$

The relative mole amounts are:
 for C, 4.483×10^{-4} mol 4.71×10^{-5} mol = 9.52
 for H, 7.094×10^{-4} mol 4.71×10^{-5} mol = 15.1
 for O, 4.71×10^{-5} mol 4.71×10^{-5} mol = 1.00

The relative mole amounts are not nice whole numbers as we would like. However, we see that if we double the relative number of moles of each compound, there are approximately 19 moles of C, 30 moles of H and 2 moles of O. If we assume these numbers are correct, the empirical formula is $C_{19}H_{30}O_2$, for which the formula weight is 290 g/mole. Since the molar mass is equal to the mass of this empirical formula, the molecular formula is the same as this empirical formula, $C_{19}H_{30}O_2$, and we have performed the analysis correctly.

In most problems where we attempt to determine an empirical formula, the relative mole amounts should work out to give a "nice" set of values for the formula. Rarely will a problem be designed that gives very odd coefficients. With experience and practice, you will recognize when a set of values is reasonable.

3.75 (a) Formula mass = 135.1 g
$$\frac{270.4 \text{ g/mol}}{135.1 \text{ g/mol}} = 2.001$$
The molecular formula is $Na_2S_4O_6$

(b) Formula mass = 73.50 g
$$\frac{147.0 \text{ g/mol}}{73.50 \text{ g/mol}} = 2.000$$
The molecular formula is $C_6H_4Cl_2$

(c) Formula mass = 60.48 g
$$\frac{181.4 \text{ g/mol}}{60.48 \text{ g/mol}} = 2.999$$
The molecular formula is $C_6H_3Cl_3$

3.77   The formula mass for the compound $C_{19}H_{30}O_2$ is 290 g/mol. Thus, the empirical and molecular formulas are equivalent.

3.79   From the information provided, we can determine the mass of mercury as the difference between the total mass and the mass of bromine:

$$\# \text{ g Hg} = 0.389 \text{ g compound} - 0.111 \text{ g Br} = 0.278 \text{ g Hg}$$

To determine the empirical formula first convert the two masses to a number of moles.

$$\# \text{ moles Hg} = \left(0.278 \text{ g Hg}\right)\left(\frac{1 \text{ mole Hg}}{200.59 \text{ g Hg}}\right) = 1.39 \times 10^{-3} \text{ moles Hg}$$

$$\# \text{ moles Br} = \left(0.111 \text{ g Br}\right)\left(\frac{1 \text{ mole Br}}{79.904 \text{ g Br}}\right) = 1.389 \times 10^{-3} \text{ moles Br}$$

Now, we would divide each of these values by the smaller quantity to determine the simplest mole ratio between the two elements. By inspection, though, we can see there are the same number of moles of Hg and Br. Consequently, the simplest mole ratio is 1:1 and the empirical formula is HgBr.

To determine the molecular formula, recall that the ratio of the molecular mass to the empirical mass is equivalent to the ratio of the molecular formula to the empirical formula. Thus, we need to calculate an empirical mass: (1 mole Hg)(200.59 g Hg/mole Hg) + (1 mole Br)(79.904 g Br/mole Br) = 280.49 g/mole HgBr. The molecular mass, as reported in the problem is 561 g/mole. The ratio of these is:

$$\frac{561 \text{ g/mole}}{280.49 \text{ g/mole}} = 2.00$$

So, the molecular formula is two times the empirical formula or Hg_2Br_2.

3.81   First, determine the amount of oxygen in the sample by subtracting the masses of the other elements from the total mass: 0.6216 g – (0.1735 g C + 0.01455 g H + 0.2024 g N) = 0.2312 g O. Now, convert these masses into a number of moles for each element:

$$\# \text{ moles C} = \left(0.1735 \text{ g C}\right)\left(\frac{1 \text{ mole C}}{12.011 \text{ g C}}\right) = 1.445 \times 10^{-2} \text{ moles C}$$

$$\# \text{ moles H} = \left(0.01455 \text{ g H}\right)\left(\frac{1 \text{ mole H}}{1.0079 \text{ g H}}\right) = 1.444 \times 10^{-2} \text{ moles H}$$

$$\# \text{ moles N} = \left(0.2024 \text{ g N}\right)\left(\frac{1 \text{ mole N}}{14.007 \text{ g N}}\right) = 1.445 \times 10^{-2} \text{ moles N}$$

$$\# \text{ moles O} = \left(0.2312 \text{ g O}\right)\left(\frac{1 \text{ mole O}}{15.999 \text{ g O}}\right) = 1.445 \times 10^{-2} \text{ moles O}$$

These are clearly all the same mole amounts, and we deduce that the empirical formula is CHNO, which has a formula weight of 43. It can be seen that the number 43 must be multiplied by the integer 3 in order to obtain the molar mass (3 × 43 = 129), and this means that the empirical formula should similarly be multiplied by 3 in order to arrive at the molecular formula, $C_3H_3N_3O_3$.

3.83   14 moles of Fe

3.85 $4Fe(s) + 3O_2(g) \rightarrow 2Fe_2O_3(s)$

3.87   (a) $Ca(OH)_2 + 2HCl \rightarrow CaCl_2 + 2H_2O$
        (b) $2AgNO_3 + CaCl_2 \rightarrow Ca(NO_3)_2 + 2AgCl$
        (c) $2Fe_2O_3 + 3C \rightarrow 4Fe + 3CO_2$
        (d) $2NaHCO_3 + H_2SO_4 \rightarrow Na_2SO_4 + 2H_2O + 2CO_2$
        (e) $2C_4H_{10} + 13O_2 \rightarrow 8CO_2 + 10H_2O$

3.89   (a) $Mg(OH)_2 + 2HBr \rightarrow MgBr_2 + 2H_2O$
        (b) $2HCl + Ca(OH)_2 \rightarrow CaCl_2 + 2H_2O$
        (c) $Al_2O_3 + 3H_2SO_4 \rightarrow Al_2(SO_4)_3 + 3H_2O$
        (d) $2KHCO_3 + H_3PO_4 \rightarrow K_2HPO_4 + 2H_2O + 2CO_2$
        (e) $C_9H_{20} + 14O_2 \rightarrow 9CO_2 + 10H_2O$

3.91   (a)

$$\text{\# moles } Na_2S_2O_3 = (0.12 \text{ moles } Cl_2)\left(\frac{1 \text{ mole } Na_2S_2O_3}{4 \text{ mole } Cl_2}\right)$$

$$= 0.030 \text{ moles } Na_2S_2O_3$$

        (b) $\text{\# moles HCl} = (0.12 \text{ moles } Cl_2)\left(\dfrac{8 \text{ mole HCl}}{4 \text{ mole } Cl_2}\right) = 0.24 \text{ moles HCl}$

        (c) $\text{\# moles } H_2O = (0.12 \text{ moles } Cl_2)\left(\dfrac{5 \text{ mole } H_2O}{4 \text{ mole } Cl_2}\right) = 0.15 \text{ moles } H_2O$

        (d) $\text{\# moles } H_2O = (0.24 \text{ moles HCl})\left(\dfrac{5 \text{ mole } H_2O}{8 \text{ mole HCl}}\right) = 0.15 \text{ moles } H_2O$

3.93   (a) $4P + 5O_2 \rightarrow P_4O_{10}$

        (b) $\text{\# g } O_2 = (6.85 \text{ g P})\left(\dfrac{1 \text{ mole P}}{30.97 \text{ g P}}\right)\left(\dfrac{5 \text{ mole } O_2}{4 \text{ mole P}}\right)\left(\dfrac{32.0 \text{ g } O_2}{1 \text{ mol } O_2}\right) = 8.85 \text{ g } O_2$

        (c) $\text{\# g } P_4O_{10} = (8.00 \text{ g } O_2)\left(\dfrac{1 \text{ mole } O_2}{32.00 \text{ g } O_2}\right)\left(\dfrac{1 \text{ mole } P_4O_{10}}{5 \text{ mole } O_2}\right)\left(\dfrac{283.9 \text{ g } P_4O_{10}}{1 \text{ mole } P_4O_{10}}\right) = 14.2 \text{ g } P_4O_{10}$

        (d) $\text{\# g P} = (7.46 \text{ g } P_4O_{10})\left(\dfrac{1 \text{ mole } P_4O_{10}}{283.9 \text{ g } P_4O_{10}}\right)\left(\dfrac{4 \text{ mole P}}{1 \text{ mole } P_4O_{10}}\right)\left(\dfrac{30.97 \text{ g P}}{1 \text{ mole P}}\right) = 3.26 \text{ g P}$

3.95 $\text{\# g of } HNO_3 = (11.45 \text{ g Cu})\left(\dfrac{1 \text{ mole Cu}}{63.546 \text{ g Cu}}\right)\left(\dfrac{8 \text{ moles } HNO_3}{3 \text{ moles Cu}}\right)\left(\dfrac{63.08 \text{ g } HNO_3}{1 \text{ mole } HNO_3}\right)$

$$= 30.31 \text{ g } HNO_3$$

3.97   (a)   First determine the amount of Fe_2O_3 that would be required to react completely with the given amount of Al:

$$\text{\# moles } Fe_2O_3 = (4.20 \text{ moles Al})\left(\frac{1 \text{ mole } Fe_2O_3}{2 \text{ moles Al}}\right) = 2.10 \text{ moles } Fe_2O_3$$

Since only 1.75 mol of Fe_2O_3 are supplied, it is the limiting reactant. This can be confirmed by calculating the amount of Al that would be required to react completely with all of the available Fe_2O_3:

$$\text{\# moles Al} = (1.75 \text{ moles Fe}_2\text{O}_3)\left(\frac{2 \text{ moles Al}}{1 \text{ mole Fe}_2\text{O}_3}\right) = 3.50 \text{ moles Al}$$

Since an excess (4.20 mol – 3.50 mol = 0.70 mol) of Al is present, Fe_2O_3 must be the limiting reactant, as determined above.

(b) $$\text{\# g Fe} = (1.75 \text{ moles Fe}_2\text{O}_3)\left(\frac{2 \text{ moles Fe}}{1 \text{ mole Fe}_2\text{O}_3}\right)\left(\frac{55.847 \text{ g Fe}}{1 \text{ mole Fe}}\right) = 195 \text{ g Fe}$$

3.99 $3AgNO_3 + FeCl_3 \rightarrow 3AgCl + Fe(NO_3)_3$
Calculate the amount of $FeCl_3$ that are required to react completely with all of the available silver nitrate:

$$\text{\# g FeCl}_3 = (18.0 \text{ g AgNO}_3)\left(\frac{1 \text{ mole AgNO}_3}{169.87 \text{ g AgNO}_3}\right)$$
$$\times \left(\frac{1 \text{ mole FeCl}_3}{3 \text{ mole AgNO}_3}\right)\left(\frac{162.21 \text{ g FeCl}_3}{1 \text{ mole FeCl}_3}\right)$$
$$= 5.73 \text{ g FeCl}_3$$

Since more than this minimum amount is available, $FeCl_3$ is present in excess, and $AgNO_3$ must be the limiting reactant.

We know that only 5.73 g $FeCl_3$ will be used. Therefore, the amount left unused is: 32.4 g total – 5.73 g used = 26.7 g $FeCl_3$

3.101   First determine the theoretical yield:

$$\text{\# g BaSO}_4 = (75.00 \text{ g Ba(NO}_3)_2)\left(\frac{1 \text{ mole Ba(NO}_3)_2}{261.34 \text{ g Ba(NO}_3)_2}\right)$$
$$\times \left(\frac{1 \text{ mole BaSO}_4}{1 \text{ mole Ba(NO}_3)_2}\right)\left(\frac{233.39 \text{ g BaSO}_4}{1 \text{ mole BaSO}_4}\right)$$
$$= 66.98 \text{ g BaSO}_4$$

Then calculate a % yield:
$$\text{\% yield} = \frac{\text{actual yield}}{\text{theoretical yield}} \times 100 = \frac{63.45 \text{ g}}{66.98 \text{ g}} \times 100 = 94.73\%$$

3.103   First, determine how much H_2SO_4 is needed to completely react with the $AlCl_3$
$$\text{\# g H}_2\text{SO}_4 = (25.00 \text{ g AlCl}_3)\left(\frac{1 \text{ mole AlCl}_3}{133.34 \text{ g AlCl}_3}\right)$$
$$\times \left(\frac{3 \text{ mole H}_2\text{SO}_4}{2 \text{ mole AlCl}_3}\right)\left(\frac{98.08 \text{ g H}_2\text{SO}_4}{1 \text{ mole H}_2\text{SO}_4}\right)$$
$$= 27.58 \text{ g H}_2\text{SO}_4$$
There is an excess of H_2SO_4 present.

Determine the theoretical yield:

$$\# g \, Al_2(SO_4)_3 = (25.00 \, g \, AlCl_3)\left(\frac{1 \, mole \, AlCl_3}{133.34 \, g \, AlCl_3}\right)$$

$$\times \left(\frac{1 \, mole \, Al_2(SO_4)_3}{2 \, mole \, AlCl_3}\right)\left(\frac{342.15 \, g \, Al_2(SO_4)_3}{1 \, mole \, Al_2(SO_4)_3}\right)$$

$$= 32.07 \, g \, Al_2(SO_4)_3$$

$$\% \, yield = \frac{actual \, yield}{theoretical \, yield} \times 100 = \frac{28.46 \, g}{32.07 \, g} \times 100 = 88.74 \, \%$$

3.105   If the yield for this reaction is only 71 % and we need to have 11.5 g of product, we will attempt to make 16 g of product. This is determined by dividing the actual yield by the percent yield. Recall that;

$$\% \, yield = \frac{actual \, yield}{theoretical \, yield} \times 100 . \text{ If we rearrange this equation we can see that}$$

$$theoretical \, yield = \frac{actual \, yield}{\% \, yield} \times 100 . \text{ Substituting the values from this problem gives the 16 g of product}$$

mentioned above.

$$\# g \, C_7H_8 = (16 \, g \, KC_7H_5O_2)\left(\frac{1 \, mole \, KC_7H_5O_2}{160.21 \, g \, KC_7H_5O_2}\right)$$

$$\times \left(\frac{1 \, mole \, C_7H_8}{1 \, mole \, KC_7H_5O_2}\right)\left(\frac{92.14 \, g \, C_7H_8}{1 \, mole \, C_7H_8}\right)$$

$$= 9.2 \, g \, C_7H_8$$

3.107 $\dfrac{0.25 \, mol \, HCl}{1 \, L \, HCl \, soln}$ and $\dfrac{1 \, L \, HCl \, soln}{0.25 \, mol \, HCl}$

3.109   For each of the following recall that molarity is defined as moles of solute divided by liters of solution.

(a)

$$M \, NaOH = \left(\frac{4.00 \, g \, NaOH}{100.0 \, mL \, NaOH \, soln}\right)\left(\frac{1 \, mol \, NaOH}{40.00 \, g \, NaOH}\right)$$

$$\times \left(\frac{1000 \, mL \, NaOH \, soln}{1 \, L \, NaOH \, soln}\right)$$

$$= 1.00 \, M \, NaOH$$

(b)

$$M \, CaCl_2 = \left(\frac{16.0 \, g \, CaCl_2}{250.0 \, mL \, CaCl_2 \, soln}\right)\left(\frac{1 \, mol \, CaCl_2}{110.98 \, g \, CaCl_2}\right)$$

$$\times \left(\frac{1000 \, mL \, CaCl_2 \, soln}{1 \, L \, CaCl_2 \, soln}\right)$$

$$= 0.577 \, M \, CaCl_2$$

(c)

$$M\,KOH = \left(\frac{14.0\,g\,KOH}{75.0\,mL\,KOH\,soln}\right)\left(\frac{1\,mol\,KOH}{56.11\,g\,KOH}\right)$$
$$\times \left(\frac{1000\,mL\,KOH\,soln}{1\,L\,KOH\,soln}\right)$$
$$= 3.33\,M\,KOH$$

(d)

$$M\,H_2C_2O_4 = \left(\frac{6.75\,g\,H_2C_2O_4}{500\,mL\,H_2C_2O_4\,soln}\right)\left(\frac{1\,mol\,H_2C_2O_4}{90.0\,g\,H_2C_2O_4}\right)$$
$$\times \left(\frac{1000\,mL\,H_2C_2O_4\,soln}{1\,L\,H_2C_2O_4\,soln}\right)$$
$$= 0.150\,M\,H_2C_2O_4$$

3.111 (a)

$$\#\,g\,NaCl = (125\,mL\,NaCl\,soln)\left(\frac{1\,L\,NaCl\,soln}{1000\,mL\,NaCl\,soln}\right)$$
$$\times \left(\frac{0.200\,mol\,NaCl}{1\,L\,NaCl\,soln}\right)\left(\frac{58.44\,g\,NaCl}{1\,mol\,NaCl}\right)$$
$$= 1.46\,g\,NaCl$$

(b)

$$\#\,g\,C_6H_{12}O_6 = (250\,mL\,C_6H_{12}O_6\,soln)\left(\frac{1\,L\,C_6H_{12}O_6\,soln}{1000\,mL\,C_6H_{12}O_6\,soln}\right)$$
$$\times \left(\frac{0.360\,mol\,C_6H_{12}O_6}{1\,L\,C_6H_{12}O_6\,soln}\right)\left(\frac{180.2\,g\,C_6H_{12}O_6}{1\,mol\,C_6H_{12}O_6}\right)$$
$$= 16.2\,g\,C_6H_{12}O_6$$

(c)

$$\#\,g\,H_2SO_4 = (250\,mL\,H_2SO_4\,soln)\left(\frac{1\,L\,H_2SO_4\,soln}{1000\,mL\,H_2SO_4\,soln}\right)$$
$$\times \left(\frac{0.250\,mol\,H_2SO_4}{1\,L\,H_2SO_4\,soln}\right)\left(\frac{98.08\,g\,H_2SO_4}{1\,mol\,H_2SO_4}\right)$$
$$= 6.12\,g\,H_2SO_4$$

3.113 $\quad M_{dil} = \dfrac{V_{concd} \bullet M_{concd}}{V_{dil}} = \dfrac{(25.0\,mL)(0.56\,M\,H_2SO_4)}{125\,mL} = 0.11\,M\,H_2SO_4$

3.115 $\quad V_{dil} = \dfrac{V_{concd} \bullet M_{concd}}{M_{dil}} = \dfrac{(25.0\,mL)(18.0\,M\,H_2SO_4)}{1.50\,M\,H_2SO_4} = 3.00\times10^2\,mL$

3.117 $\quad V_{dil} = \dfrac{V_{concd} \bullet M_{concd}}{M_{dil}} = \dfrac{(150\,mL)(2.5\,M\,KOH)}{1.0\,M\,KOH} = 380\,mL$

This is the total volume we need. In order to have this final volume we must add 230 mL (380 mL – 150 mL) of water to our original solution.

3.119

$$M\,KOH = \frac{(20.78\,\text{mL HCl soln})\left(\dfrac{1\,\text{L HCl soln}}{1000\,\text{mL HCl soln}}\right)\left(\dfrac{0.116\,\text{mol HCl}}{1\,\text{L HCl soln}}\right)\left(\dfrac{1\,\text{mol KOH}}{1\,\text{mol HCl}}\right)}{(21.34\,\text{mL KOH soln})\left(\dfrac{1\,\text{L KOH soln}}{1000\,\text{mL KOH soln}}\right)}$$

$$= 0.113\,M\,KOH$$

3.121   First, determine the number of moles of Na_2CO_3 that are to react:

$$\# \text{ mol } Na_2CO_3 = (0.0200\,\text{L } Na_2CO_3\,\text{soln})\left(\frac{0.15\,\text{mol } Na_2CO_3}{1\,\text{L } Na_2CO_3\,\text{soln}}\right)$$

$$= 3.0 \times 10^{-3}\,\text{mol } Na_2CO_3$$

Next, calculate the number of moles of $NiCl_2$ that are required based on the coefficients in the balanced equation: $NiCl_2 + Na_2CO_3 \rightarrow NiCO_3 + 2NaCl$

$$\# \text{ mol } NiCl_2 = (3.0 \times 10^{-3}\,\text{mol } Na_2CO_3)\left(\frac{1\,\text{mol } NiCl_2}{1\,\text{mol } Na_2CO_3}\right)$$

$$= 3.0 \times 10^{-3}\,\text{mol } NiCl_2$$

Next, calculate the volume of 0.25 M $NiCl_2$ solution that will contain this number of moles:

$$\# \text{ L } NiCl_2 = (3.0 \times 10^{-3}\,\text{mol } NiCl_2)\left(\frac{1\,\text{L } NiCl_2\,\text{soln}}{0.25\,\text{mol } NiCl_2}\right)$$

$$= 1.2 \times 10^{-2}\,\text{L } NiCl_2\,\text{soln}$$

$$= 12\,\text{mL } NiCl_2\,\text{soln}$$

Alternatively, we can write: $3.0 \times 10^{-3}\,\text{mol} / 0.25\,\text{mol/L} = 0.012\,\text{L}$
The number of grams of product to expect is calculated as follows:

$$\# \text{ g } NiCO_3 = (3.0 \times 10^{-3}\,\text{mol } NiCl_2)\left(\frac{1\,\text{mol } NiCO_3}{1\,\text{mol } NiCl_2}\right)\left(\frac{118.7\,\text{g } NiCO_3}{1\,\text{mol } NiCO_3}\right)$$

$$= 0.36\,\text{g } NiCO_3$$

3.123   First, determine the number of moles of H_3PO_4 that are to react:
$0.0250\,\text{L} \times 0.250\,\text{mol/L} = 6.25 \times 10^{-3}\,\text{mol } H_3PO_4$
Next, determine the number of moles of NaOH that are required by the balanced chemical equation: $H_3PO_4 + 3NaOH \rightarrow Na_3PO_4 + 3H_2O$
$6.25 \times 10^{-3}\,\text{mol } H_3PO_4 \times (3\,\text{mol NaOH}/1\,\text{mol } H_3PO_4) = 0.0188\,\text{mol NaOH}$

Last, calculate the volume of NaOH solution that will deliver this required number of moles of NaOH:
$0.0188\,\text{mol NaOH} \times 1000\,\text{mL}/0.100\,\text{mol} = 188\,\text{mL}$

Alternatively, we can write:
$0.0188\,\text{mol} / 0.100\,\text{mol/L} = 0.188\,\text{L}$

Additional Exercises

3.125

$$\text{\# yrs} = (6.02 \times 10^{23} \text{ pennies})\left(\frac{1 \text{ dollar}}{100 \text{ pennies}}\right)\left(\frac{1 \text{ second}}{5.00 \times 10^8 \text{ dollars}}\right)$$

$$\times \left(\frac{1 \text{ min}}{60 \text{ sec}}\right)\left(\frac{1 \text{ hr}}{60 \text{ min}}\right)\left(\frac{1 \text{ day}}{24 \text{ hrs}}\right)\left(\frac{1 \text{ yr}}{365 \text{ days}}\right) = 382{,}000 \text{ yrs}$$

3.127   First determine the percentage by weight of each element in the respective original samples. This is done by determining the mass of the element in question present in each of the original samples. The percentage by weight of each element in the unknown will be the same as the values we calculate.

$$\text{\# g Ca} = (0.160 \text{ g CaCO}_3)\left(\frac{1 \text{ mole CaCO}_3}{100.09 \text{ g CaCO}_3}\right)\left(\frac{1 \text{ mole Ca}}{1 \text{ mole CaCO}_3}\right)\left(\frac{40.1 \text{ g Ca}}{1 \text{ mole Ca}}\right)$$

$$= 0.0641 \text{ g Ca}$$

% Ca = $(0.0641/0.250) \times 100 = 25.6$ % Ca

$$\text{\# g S} = (0.344 \text{ g BaSO}_4)\left(\frac{1 \text{ mole BaSO}_4}{233.8 \text{ g BaSO}_4}\right)\left(\frac{1 \text{ mole S}}{1 \text{ mole BaSO}_4}\right)\left(\frac{32.07 \text{ g S}}{1 \text{ mole S}}\right)$$

$$= 0.0472 \text{ g S}$$

% S = $(0.0472/0.115) \times 100 = 41.0$ % S

$$\text{\# g N} = (0.155 \text{ g NH}_3)\left(\frac{1 \text{ mole NH}_3}{17.03 \text{ g NH}_3}\right)\left(\frac{1 \text{ mole N}}{1 \text{ mole NH}_3}\right)\left(\frac{14.01 \text{ g N}}{1 \text{ mole N}}\right)$$

$$= 0.128 \text{ g N}$$

% N = $(0.128/0.712) \times 100 = 18.0$ % N

% C = $100.0 - (25.6 + 41.0 + 18.0) = 15.4$ % C. Next, we assume 100 g of the compound, and convert these weight percentages into mole amounts:

$$\text{\# moles Ca} = (25.6 \text{ g Ca})\left(\frac{1 \text{ mole Ca}}{40.08 \text{ g Ca}}\right) = 0.639 \text{ moles Ca}$$

$$\text{\# moles S} = (41.0 \text{ g S})\left(\frac{1 \text{ mole S}}{32.07 \text{ g S}}\right) = 1.28 \text{ moles S}$$

$$\text{\# moles N} = (18.0 \text{ g N})\left(\frac{1 \text{ mole N}}{14.07 \text{ g N}}\right) = 1.28 \text{ moles N}$$

$$\text{\# moles C} = (15.4 \text{ g C})\left(\frac{1 \text{ mole C}}{12.01 \text{ g C}}\right) = 1.28 \text{ moles C}$$

Dividing each of these mole amounts by the smallest, we have:

For Ca: 0.639 mol / 0.639 mol = 1.00
For S: 1.28 mol / 0.639 mol = 2.00
For N: 1.28 mol / 0.639 mol = 2.00
For C: 1.28 mol / 0.639 mol = 2.00

The empirical formula is therefore $CaC_2S_2N_2$, and the mass of the empirical unit is Ca + 2S + 2N + 2C = 156 g/mol. Since the molecular mass is the same as the empirical mass, the molecular formula is $CaC_2S_2N_2$.

3.129

$$\# \text{ g } (NH_2)_2CO = (6.00 \text{ g N}) \left(\frac{1 \text{ mole N}}{14.007 \text{ g N}} \right) \left(\frac{1 \text{ mole } (NH_2)_2CO}{2 \text{ moles N}} \right) \left(\frac{60.06 \text{ g } (NH_2)_2CO}{1 \text{ mole } (NH_2)_2CO} \right)$$

$$= 12.9 \text{ g } (NH_2)_2CO$$

Assume the hydrogen is the limiting reactant.

$$\text{lb O}_2 = (227,641 \text{ lb H}_2) \left(\frac{453.59237 \text{ g}}{1 \text{ lb}} \right) \left(\frac{1 \text{ mol H}_2}{2.016 \text{ g H}_2} \right) \left(\frac{1 \text{ mol O}_2}{2 \text{ mol H}_2} \right)$$

$$\times \left(\frac{32.0 \text{ g O}_2}{1 \text{ mol O}_2} \right) \left(\frac{1 \text{ lb O}_2}{453.59237 \text{ g O}_2} \right) = 1,806,675 \text{ lb O}_2$$

3.131 Assume the hydrogen is the limiting reactant.

$$\text{lb O}_2 = (227,641 \text{ lb H}_2) \left(\frac{453.59237 \text{ g}}{1 \text{ lb}} \right) \left(\frac{1 \text{ mol H}_2}{2.01588 \text{ g H}_2} \right) \left(\frac{1 \text{ mol O}_2}{2 \text{ mol H}_2} \right)$$

$$\times \left(\frac{31.9988 \text{ g O}_2}{1 \text{ mol O}_2} \right) \left(\frac{1 \text{ lb O}_2}{453.59237 \text{ g O}_2} \right) = 1,806,714 \text{ lb O}_2$$

Since this is more than the amount of O_2 that is supplied, the limiting reactant must be O_2. Next calculate the amount of H_2 needed to react completely with all of the available O_2.

$$\text{lb H}_2 = (1,361,936 \text{ lb O}_2) \left(\frac{453.59237 \text{ g}}{1 \text{ lb}} \right) \left(\frac{1 \text{ mol O}_2}{31.9988 \text{ g O}_2} \right) \left(\frac{2 \text{ mol H}_2}{1 \text{ mol O}_2} \right)$$

$$\times \left(\frac{2.01588 \text{ g H}_2}{1 \text{ mol H}_2} \right) \left(\frac{1 \text{ lb H}_2}{453.59237 \text{ g H}_2} \right) = 171,600 \text{ lb H}_2$$

Since only 171,600 lb. of H_2 reacted, there are 227,641 lb. – 171,600 lb. = 56,041 lb. of unreacted H_2.

3.133 In order to solve this problem we must recall two things; molarity is defined as the number of moles of solute per liter of solution and that molarity times volume equals the number of moles. Set up the following algebraic equation:

$$\frac{(50.0 \text{ mL})(0.40 \text{ M}) + (x \text{ mL})(0.10 \text{ M})}{(50.0 \text{ mL} + x \text{ mL})} = 0.25 \text{ M}$$

The numerator of this expression is the total number of moles of solution and the denominator is the total volume of the solution. We can now solve this equation for x and we find that x = 5.0×10^1 mL.

Practice Exercises

P4.1   (a) $MgCl_2(s) \rightarrow Mg^{2+}(aq) + 2Cl^-(aq)$
       (b) $Al(NO_3)_3(s) \rightarrow Al^{3+}(aq) + 3NO_3^-(aq)$
       (c) $Na_2CO_3(s) \rightarrow 2Na^+(aq) + CO_3^{2-}(aq)$
       (d) $(NH_4)_2SO_4(s) \rightarrow 2NH_4^+(aq) + SO_4^{2-}(aq)$

P4.2   molecular: $CdCl_2(aq) + Na_2S(aq) \rightarrow CdS(s) + 2NaCl(aq)$
       ionic: $Cd^{2+}(aq) + 2Cl^-(aq) + 2Na^+(aq) + S^{2-}(aq) \rightarrow CdS(s) + 2Na^+(aq) + 2Cl^-(aq)$
       net ionic: $Cd^{2+}(aq) + S^{2-}(aq) \rightarrow CdS(s)$

P4.3   (a) molecular: $AgNO_3(aq) + NH_4Cl(aq) \rightarrow AgCl(s) + NH_4NO_3(aq)$
       ionic: $Ag^+(aq) + NO_3^-(aq) + NH_4^+(aq) + Cl^-(aq) \rightarrow AgCl(s) + NH_4^+(aq) + NO_3^-(aq)$
       net ionic: $Ag^+(aq) + Cl^-(aq) \rightarrow AgCl(s)$

       (b) molecular: $Na_2S(aq) + Pb(C_2H_3O_2)_2 \rightarrow 2NaC_2H_3O(aq) + PbS(s)$
       ionic: $2Na^+(aq) + S^{2-}(aq) + Pb^{2+}(aq) + 2C_2H_3O^-(aq) \rightarrow 2Na^+(aq) + 2C_2H_3O^-(aq) + PbS(s)$
       net ionic: $S^{2-}(aq) + Pb^{2+}(aq) \rightarrow PbS(s)$

P4.4   (a) $HCHO_2(\ell) + H_2O \rightarrow H_3O^+(aq) + CHO_2^-(aq)$
       (b) $H_3PO_4(\ell) + H_2O \rightarrow H_3O^+(aq) + H_2PO_4^-(aq)$
               $H_2PO_4^-(aq) + H_2O \rightarrow H_3O^+(aq) + HPO_4^{2-}(aq)$
               $HPO_4^{2-}(aq) + H_2O \rightarrow H_3O^+(aq) + PO_4^{3-}(aq)$

P4.5 $HNO_2(aq) + H_2O \rightleftharpoons H_3O^+(aq) + NO_2^-(aq)$

P4.6   molecular: $2HCl(aq) + Ca(OH)_2(aq) \rightarrow CaCl_2(aq) + 2H_2O(\ell)$
       ionic: $2H^+(aq) + 2Cl^-(aq) + Ca^{2+}(aq) + 2OH^-(aq) \rightarrow Ca^{2+}(aq) + 2Cl^-(aq) + 2H_2O(\ell)$
       net ionic: $2H^+(aq) + 2OH^-(aq) \rightarrow 2H_2O(\ell)$

P4.7   (a)   molecular: $HCl(aq) + KOH(aq) \rightarrow H_2O(\ell) + KCl(aq)$
               ionic: $H^+(aq) + Cl^-(aq) + K^+(aq) + OH^-(aq)$
                              $\rightarrow H_2O(\ell) + K^+(aq) + Cl^-(aq)$
               net ionic: $H^+(aq) + OH^-(aq) \rightarrow H_2O(\ell)$

       (b)   molecular: $HCHO_2(aq) + LiOH(aq) \rightarrow H_2O(\ell) + LiCHO_2(aq)$
               ionic: $HCHO_2(aq) + Li^+(aq) + OH^-(aq)$
                              $\rightarrow H_2O(\ell) + Li^+(aq) + CHO_2^-(aq)$
               net ionic: $HCHO_2(aq) + OH^-(aq) \rightarrow H_2O(\ell) + CHO_2^-(aq)$

       (c)   molecular: $N_2H_4(aq) + HCl(aq) \rightarrow N_2H_5Cl(aq)$
               ionic: $N_2H_4(aq) + H^+(aq) + Cl^-(aq) \rightarrow N_2H_5^+(aq) + Cl^-(aq)$
               net ionic: $N_2H_4(aq) + H^+(aq) \rightarrow N_2H_5^+(aq)$

P4.8   molecular: $Al(OH)_3(s) + 3HCl(aq) \rightarrow AlCl_3(aq) + 3H_2O(\ell)$
       ionic: $Al(OH)_3(s) + 3H^+(aq) + 3Cl^-(aq) \rightarrow Al^{3+}(aq) + 3Cl^-(aq) + 3H_2O(\ell)$
       net ionic: $Al(OH)_3(s) + 3H^+(aq) \rightarrow Al^{3+}(aq) + 3H_2O(\ell)$

P4.9   (a)   Formic acid, a weak acid will form.
             Net ionic equation: $H^+(aq) + CHO_2^-(aq) \rightleftharpoons HCHO_2(aq)$

(b)   Carbonic acid will form and it will further dissociate to water and carbon dioxide.
$$CuCO_3(s) + 2H^+(aq) \rightarrow 2CO_2(g) + 2H_2O(\ell) + Cu^{2+}(aq)$$

(c)   NR

(d)   Insoluble nickel hydroxide will precipitate.
$$Ni^{2+}(aq) + 2OH^-(aq) \rightarrow Ni(OH)_2(s)$$

P4.10 $FeCl_3 \rightarrow Fe^{3+} + 3Cl^-$

$$M\,Fe^{3+} = \left(\frac{0.40\ mol\ FeCl_3}{1\,L\ FeCl_3\ soln}\right)\left(\frac{1\,mol\ Fe^{3+}}{1\,mol\ FeCl_3}\right) = 0.40\ M\ Fe^{3+}$$

$$M\,Cl^- = \left(\frac{0.40\ mol\ FeCl_3}{1\,L\ FeCl_3\ soln}\right)\left(\frac{3\,mol\ Cl^-}{1\,mol\ FeCl_3}\right) = 1.2\ M\ Cl^-$$

P4.11 $$M\,Na^+ = \left(\frac{0.250\ mol\ PO_4^{3-}}{1\,L\ Na_3PO_4\ soln}\right)\left(\frac{3\,mol\ Na^+}{1\,mol\ PO_4^{3-}}\right) = 0.750\ M\ PO_4^{3-}$$

P4.12   The balanced net ionic equation is: $Fe^{2+}(aq) + 2OH^-(aq) \rightarrow Fe(OH)_2(s)$.
First determine the number of moles of Fe^{2+} present.

$$\#\ moles\ Fe^{2+} = \left(60.0\ mL\ FeCl_2\ solution\right)\left(\frac{0.250\ mol\ FeCl_2}{1000\ mL\ solution}\right)\left(\frac{1\,mol\ Fe^{2+}}{1\,mol\ FeCl_2}\right)$$

$$= 1.50 \times 10^{-2}\ mol\ Fe^{2+}$$

Now, determine the amount of KOH needed to react with the Fe^{2+}.

$$\#\ mL\ KOH = \left(1.50 \times 10^{-2}\ mol\ Fe^{2+}\right)\left(\frac{2\,mol\ OH^-}{1\,mol\ Fe^{2+}}\right)$$

$$\times \left(\frac{1\,mol\ KOH}{1\,mol\ OH^-}\right)\left(\frac{1000\ mL\ solution}{0.500\ mol\ KOH}\right)$$

$$= 60.0\ mL\ KOH$$

P4.13   The net ionic equation is $Ba^{2+}(aq) + SO_4^{2-}(aq) \rightarrow BaSO_4(s)$
First, determine the initial number of moles of Ba^{2+} ion that are present:

$$\#\ mol\ Ba^{2+} = (20.0\ mL\ BaCl_2\ soln)\left(\frac{0.600\ mol\ BaCl_2}{1000\ mL\ BaCl_2\ soln}\right)\left(\frac{1\,mol\ Ba^{2+}}{1\,mol\ BaCl_2}\right)$$

$$= 1.20 \times 10^{-2}\ mol\ Ba^{2+}$$

Next, determine the initial number of moles of sulfate ion that are present:

$$\#\ mol\ SO_4^{2-} = (30.0\ mL\ MgSO_4\ soln)\left(\frac{0.500\ mol\ MgSO_4}{1000\ mL\ MgSO_4\ soln}\right)\left(\frac{1\,mol\ SO_4^{2-}}{1\,mol\ MgSO_4}\right)$$

$$= 1.50 \times 10^{-2}\ mol\ SO_4^{2-}$$

Now determine the number of moles of barium ion that are required to react with this much sulphate ion, and compare the result to the amount of barium ion that is available:

$$\# \, mol \, Ba^{2+} = (1.50 \times 10^{-2} \, mol \, SO_4^{2-}) \left(\frac{1 \, mol \, Ba^{2+}}{1 \, mol \, SO_4^{2-}} \right)$$

$$= 1.50 \times 10^{-2} \, mol \, Ba^{2+}$$

Since there is not this much Ba^{2+} available according to the above calculation, then we can conclude that Ba^{2+} must be the limiting reactant, and that subsequent calculations should be based on the number of moles of it that are present:

Since this reaction is 1:1, we know that 1.20×10^{-2} mole of $BaSO_4$ will be formed.

If we assume that the $BaSO_4$ is completely insoluble, then the concentration of barium ion will be essentially zero. The concentrations of the other ions are determined as follows:

$$\# \, M \, Cl^- = \frac{(20.0 \, mL \, BaCl_2 \, soln) \left(\frac{0.600 \, mol \, BaCl_2}{1000 \, mL \, BaCl_2 \, soln} \right) \left(\frac{2 \, mol \, Cl^-}{1 \, mol \, BaCl_2} \right)}{((20.0 + 30.0) \, mL \, soln) \left(\frac{1 \, L \, soln}{1000 \, mL \, soln} \right)}$$

$$= 0.480 \, M \, Cl^-$$

$$\# \, M \, Mg^{2+} = \frac{(30.0 \, mL \, MgSO_4 \, soln) \left(\frac{0.500 \, mol \, MgSO_4}{1000 \, mL \, MgSO_4 \, soln} \right) \left(\frac{1 \, mol \, Mg^{2+}}{1 \, mol \, MgSO_4} \right)}{((30.0 + 20.0) \, mL \, soln) \left(\frac{1 \, L \, soln}{1000 \, mL \, soln} \right)}$$

$$= 0.300 \, M \, Mg^{2+}$$

For sulfate, we subtract the amount that reacted with the Ba^{2+}:
$$\# \, mol \, SO_4^{2-} = 1.50 \times 10^{-2} \, mol - 1.20 \times 10^{-2} \, mol = 3.0 \times 10^{-3} \, mol$$

This allows a calculation of the final sulfate concentration:

$$\# \, M \, SO_4^{2-} = \frac{3.0 \times 10^{-3} \, mol \, SO_4^{2-}}{((30.0 + 20.0) \, mL \, soln) \left(\frac{1 \, L \, soln}{1000 \, mL \, soln} \right)}$$

$$= 6.0 \times 10^{-2} \, M \, SO_4^{2-}$$

P4.14   (a)
$$\# \, mol \, Ca^{2+} = (0.736 \, g \, CaSO_4) \left(\frac{1 \, mol \, CaSO_4}{136.14 \, g \, CaSO_4} \right) \left(\frac{1 \, mol \, Ca^{2+}}{1 \, mol \, CaSO4} \right)$$

$$= 5.41 \times 10^{-3} \, mol \, Ca^{2+}$$

(b)   Since all of the Ca^{2+} is precipitated as $CaSO_4$, there were originally 5.41×10^{-3} moles of Ca^{2+} in the sample.

(c)   All of the Ca^{2+} comes from $CaCl_2$, so there were 5.41×10^{-3} moles of $CaCl_2$ in the sample.

(d)
$$\# \text{ g CaCl2} = \left(5.41 \times 10^{-3} \text{ mol CaCl}_2\right)\left(\frac{110.98 \text{ g CaCl}_2}{1 \text{ mol CaCl}_2}\right)$$
$$= 0.600 \text{ g CaCl}_2$$

(e)
$$\% \text{ CaCl}_2 = \frac{0.600 \text{ g CaCl}_2}{2.000 \text{ g sample}} \times 100 = 30.0 \% \text{ CaCl}_2$$

P4.15

$$M \text{ H}_2\text{SO}_4 = \left[\frac{(36.42 \text{ mL NaOH soln})\left(\frac{1 \text{ L NaOH soln}}{1000 \text{ mL NaOH soln}}\right)\left(\frac{0.147 \text{ mol NaOH}}{1 \text{ L NaOH soln}}\right)\left(\frac{1 \text{ mol H}_2\text{SO}_4}{2 \text{ mol NaOH}}\right)}{(15.00 \text{ mL H}_2\text{SO}_4 \text{ soln})\left(\frac{1 \text{ L H}_2\text{SO}_4 \text{ soln}}{1000 \text{ mL H}_2\text{SO}_4 \text{ soln}}\right)}\right]$$

$$= 0.178 \text{ M H}_2\text{SO}_4$$

P4.16

$$M \text{ HCl} =$$

$$\left[\frac{(11.00 \text{ mL KOH soln})\left(\frac{1 \text{ L KOH soln}}{1000 \text{ mL NaOH soln}}\right)\left(\frac{0.0100 \text{ mol KOH}}{1 \text{ L KOH soln}}\right)\left(\frac{1 \text{ mol HCl}}{1 \text{ mol KOH}}\right)}{(5.00 \text{ mL HCl soln})\left(\frac{1 \text{ L HCl soln}}{1000 \text{ mL HCl soln}}\right)}\right]$$

$$= 0.0220 \text{ M HCl}$$

$$\# \text{ g HCl} = (5.00 \text{ mL HCl})\left(\frac{0.0220 \text{ mol HCl}}{1000 \text{ mL HCl}}\right)\left(\frac{36.5 \text{ g HCl}}{1 \text{ mol HCl}}\right) = 4.02 \times 10^{-3} \text{ g HCl}$$

$$\text{weight } \% = \frac{4.02 \times 10^{-3} \text{ g}}{5.00 \text{ g}} \times 100\% = 0.0803\%$$

Review Problems

4.25 (a) $\text{LiCl(s)} \rightarrow \text{Li}^+\text{(aq)} + \text{Cl}^-\text{(aq)}$

    (b) $\text{BaCl}_2\text{(s)} \rightarrow \text{Ba}^{2+}\text{(aq)} + 2\text{Cl}^-\text{(aq)}$

    (c) $\text{Al(C}_2\text{H}_3\text{O}_2)_3\text{(s)} \rightarrow \text{Al}^{3+}\text{(aq)} + 3\text{C}_2\text{H}_3\text{O}_2^-\text{(aq)}$

    (d) $\text{(NH}_4)_2\text{CO}_3\text{(s)} \rightarrow 2\text{NH}_4^+\text{(aq)} + \text{CO}_3^{2-}\text{(aq)}$

    (e) $\text{FeCl}_3\text{(s)} \rightarrow \text{Fe}^{3+}\text{(aq)} + 3\text{Cl}^-\text{(aq)}$

4.27 (a) ionic: $2\text{NH}_4^+\text{(aq)} + \text{CO}_3^{2-}\text{(aq)} + \text{Mg}^{2+}\text{(aq)} + 2\text{Cl}^-\text{(aq)} \rightarrow$
                $2\text{NH}_4^+\text{(aq)} + 2\text{Cl}^-\text{(aq)} + \text{MgCO}_3\text{(s)}$
        net: $\text{Mg}^{2+}\text{(aq)} + \text{CO}_3^{2-}\text{(aq)} \rightarrow \text{MgCO}_3\text{(s)}$

    (b) ionic: $\text{Cu}^{2+}\text{(aq)} + 2\text{Cl}^-\text{(aq)} + 2\text{Na}^+\text{(aq)} + 2\text{OH}^-\text{(aq)} \rightarrow$
                $\text{Cu(OH)}_2\text{(s)} + 2\text{Na}^+\text{(aq)} + 2\text{Cl}^-\text{(aq)}$
        net: $\text{Cu}^{2+}\text{(aq)} + 2\text{OH}^-\text{(aq)} \rightarrow \text{Cu(OH)}_2\text{(s)}$

    (c) ionic: $3\text{Fe}^{2+}\text{(aq)} + 3\text{SO}_4^{2-}\text{(aq)} + 6\text{Na}^+\text{(aq)} + 2\text{PO}_4^{3-}\text{(aq)} \rightarrow$
                $\text{Fe}_3\text{(PO}_4)_2\text{(s)} + 6\text{Na}^+\text{(aq)} + 3\text{SO}_4^{2-}\text{(aq)}$
        net: $3\text{Fe}^{2+}\text{(aq)} + 2\text{PO}_4^{3-}\text{(aq)} \rightarrow \text{Fe}_3\text{(PO}_4)_2\text{(s)}$

(d) ionic: $2Ag^+(aq) + 2C_2H_3O_2^-(aq) + Ni^{2+}(aq) + 2Cl^-(aq) \rightarrow$
$2AgCl(s) + Ni^{2+}(aq) + 2C_2H_3O_2^-(aq)$

net: $2Ag^+(aq) + 2Cl^-(aq) \rightarrow 2AgCl(s)$

4.29 $Cu^{2+}(aq) + S^{2-}(aq) \rightarrow CuS(s)$

4.31 molecular: $AgNO_3(aq) + NaBr(aq) \rightarrow AgBr(s) + NaNO_3(aq)$
ionic: $Ag^+(aq) + NO_3^-(aq) + Na^+(aq) + Br^-(aq) \rightarrow AgBr(s) + Na^+(aq) + NO_3^-(aq)$
net: $Ag^+(aq) + Br^-(aq) \rightarrow AgBr(s)$

4.33 This is an ionization reaction: $HClO_4(\ell) + H_2O(\ell) \rightleftharpoons H_3O^+(aq) + ClO_4^-(aq)$

4.35 $N_2H_4(aq) + H_2O(\ell) \rightleftharpoons N_2H_5^+(aq) + OH^-(aq)$

4.37 These are not reversible reactions, that is the reverse reaction has practically no tendency to occur.

4.39 $H_2CO_3(aq) + H_2O(\ell) \rightleftharpoons H_3O^+(aq) + HCO_3^-(aq)$
$HCO_3^-(aq) + H_2O(\ell) \rightleftharpoons H_3O^+(aq) + CO_3^{2-}(aq)$

4.41 (a) molecular: $Ca(OH)_2(aq) + 2HNO_3(aq) \rightarrow Ca(NO_3)_2(aq) + 2H_2O$
ionic: $Ca^{2+}(aq) + 2OH^-(aq) + 2H^+(aq) + 2NO_3^-(aq) \rightarrow$
$Ca^{2+}(aq) + 2NO_3^-(aq) + 2H_2O$
net: $H^+(aq) + OH^-(aq) \rightarrow H_2O$

(b) molecular: $Al_2O_3(s) + 6HCl(aq) \rightarrow 2AlCl_3(aq) + 3H_2O$
ionic: $Al_2O_3(s) + 6H^+(aq) + 6Cl^-(aq) \rightarrow$
$2Al^{3+}(aq) + 6Cl^-(aq) + 3H_2O$
net: $Al_2O_3(s) + 6H^+(aq) \rightarrow 2Al^{3+}(aq) + 3H_2O$

(c) molecular: $Zn(OH)_2(s) + H_2SO_4(aq) \rightarrow ZnSO_4(aq) + 2H_2O$
ionic: $Zn(OH)_2(s) + 2H^+(aq) + SO_4^{2-}(aq) \rightarrow$
$Zn^{2+}(aq) + SO_4^{2-}(aq) + 2H_2O$
net: $Zn(OH)_2(s) + 2H^+(aq) \rightarrow Zn^{2+}(aq) + 2H_2O$

4.43 (a) $2H^+(aq) + CO_3^{2-}(aq) \rightarrow H_2O(\ell) + CO_2(g)$
(b) $NH_4^+(aq) + OH^-(aq) \rightarrow NH_3(aq) + H_2O(\ell)$

4.45 These reactions have the following "driving forces":
(a) formation of insoluble $Cr(OH)_3$
(b) formation of water, a weak electrolyte

4.47 The soluble ones are (a), (b), and (d).

4.49 The insoluble ones are (a), (d), and (f).

4.51 a) $3HNO_3(aq) + Cr(OH)_3(s) \rightarrow Cr(NO_3)_3(aq) + 3H_2O(\ell)$
ionic: $3H^+(aq) + 3NO_3^-(aq) + Cr(OH)_3(s) \rightarrow$
$Cr^{3+}(aq) + 3NO_3^-(aq) + 3H_2O(\ell)$
net: $3H^+(aq) + Cr(OH)_3(s) \rightarrow Cr^{3+}(aq) + 3H_2O(\ell)$

(b) $HClO_4(aq) + NaOH(aq) \rightarrow NaClO_4(aq) + H_2O(\ell)$

ionic: $H^+(aq) + ClO_4^-(aq) + Na^+(aq) + OH^-(aq) \rightarrow$
$$Na^+(aq) + ClO_4^-(aq) + H_2O(\ell)$$

net: $H^+(aq) + OH^-(aq) \rightarrow H_2O(\ell)$

(c) $Cu(OH)_2(s) + 2HC_2H_3O_2(aq) \rightarrow Cu(C_2H_3O_2)_2(aq) + 2H_2O(\ell)$

ionic: $Cu(OH)_2(s) + 2H^+ + 2C_2H_3O_2^-(aq) \rightarrow$
$$Cu^{2+}(aq) + 2C_2H_3O_2^-(aq) + 2H_2O(\ell)$$

net: $Cu(OH)_2(s) + 2H^+(aq) \rightarrow Cu^{2+}(aq) + 2H_2O(\ell)$

(d) $ZnO(s) + 2HBr(aq) \rightarrow ZnBr_2(aq) + H_2O(\ell)$

ionic: $ZnO(s) + 2H^+(aq) + 2Br^-(aq) \rightarrow Zn^{2+}(aq) + 2Br^-(aq) + H_2O(\ell)$

net: $ZnO(s) + 2H^+(aq) \rightarrow Zn^{2+}(aq) + H_2O(\ell)$

4.53   (a) $Na_2SO_3(aq) + Ba(NO_3)_2(aq) \rightarrow BaSO_3(s) + 2NaNO_3(aq)$

ionic: $2Na^+(aq) + SO_3^{2-}(aq) + Ba^{2+}(aq) + 2NO_3^-(aq) \rightarrow$
$$BaSO_3(s) + 2Na^+(aq) + 2NO_3^-(aq)$$

net: $Ba^{2+}(aq) + SO_3^{2-}(aq) \rightarrow BaSO_3(s)$

(b) $K_2S(aq) + ZnCl_2(aq) \rightarrow ZnS(s) + 2KCl(aq)$

ionic: $2K^+(aq) + S^{2-}(aq) + Zn^{2+}(aq) + 2Cl^-(aq)$
$$ZnS(s) + 2K^+(aq) + 2Cl^-(aq)$$

net: $Zn^{2+}(aq) + S^{2-}(aq) \rightarrow ZnS(s)$

(c) $2NH_4Br(aq) + Pb(C_2H_3O_2)_2(aq) \rightarrow 2NH_4C_2H_3O_2(aq) + PbBr_2(s)$

ionic: $2NH_4^+(aq) + 2Br^-(aq) + Pb^{2+}(aq) + 2C_2H_3O_2^-(aq) \rightarrow$
$$2NH_4^+(aq) + 2C_2H_3O_2^-(aq) + PbBr_2(s)$$

net: $Pb^{2+}(aq) + 2Br^-(aq) \rightarrow PbBr_2(s)$

(d) $2NH_4ClO_4(aq) + Cu(NO_3)_2(aq) \rightarrow Cu(ClO_4)_2(aq) + 2NH_4NO_3(aq)$

ionic: $2NH_4^+(aq) + 2ClO_4^-(aq) + Cu^{2+}(aq) + 2NO_3^-(aq) \rightarrow$
$$Cu^{2+}(aq) + 2ClO_4^-(aq) + 2NO_3^-(aq) + 2NH_4^+(aq)$$

net: N.R.

4.55   There are numerous possible answers. One of many possible sets of answers would be:

(a) $NaHCO_3(aq) + HCl(aq) \rightarrow NaCl(aq) + CO_2(g) + H_2O(\ell)$

(b) $FeCl_2(aq) + 2NaOH(aq) \rightarrow Fe(OH)_2(s) + 2NaCl(aq)$

(c) $Ba(NO_3)_2(aq) + K_2SO_3(aq) \rightarrow BaSO_3(s) + 2KNO_3(aq)$

(d) $2AgNO_3(aq) + Na_2S(aq) \rightarrow Ag_2S(s) + 2NaNO_3(aq)$

(e) $ZnO(s) + 2HCl(aq) \rightarrow ZnCl_2(aq) + H_2O(\ell)$

4.57   (a) $KOH \rightarrow K^+ + OH^-$

1.25 mol/L x 0.0350 L = 0.0438 mol KOH

0.0438 mol KOH × 1 mol OH^-/mol KOH = 0.0438 mol OH^-

0.0438 mol KOH × 1 mol K^+/mol KOH = 0.0438 mol K^+

(b) $CaCl_2 \rightarrow Ca^{2+} + 2Cl^-$

0.45 mol/L x× 0.0323 L = 0.015 mol $CaCl_2$

0.015 mol $CaCl_2$ × 1 mol Ca^{2+}/mol $CaCl_2$ = 0.015 mol Ca^{2+}

0.015 mol $CaCl_2$ × 2 mol Cl^-/mol $CaCl_2$ = 0.030 mol Cl^-

(c) $AlCl_3 \rightarrow Al^{3+} + 3Cl^-$

$$\text{\# moles } AlCl_3 = (50 \text{ mL } AlCl_3)\left(\frac{0.40 \text{ mol } AlCl_3}{1000 \text{ mL}}\right) = 2.0 \times 10^{-2} \text{ moles } AlCl_3$$

$$\text{\# moles } Al^{3+} = (2.0 \times 10^{-2} \text{ mol } AlCl_3)\left(\frac{1 \text{ mol } Al^{3+}}{1 \text{ mol } AlCl_3}\right) = 2.0 \times 10^{-2} \text{ moles } Al^{3+}$$

$$\text{\# moles } Cl^- = (2.0 \times 10^{-2} \text{ mol } AlCl_3)\left(\frac{3 \text{ mol } Cl^-}{1 \text{ mol } AlCl_3}\right) = 6.0 \times 10^{-2} \text{ moles } Cl^-$$

4.59 (a) $Cr(NO_3)_2 \rightarrow Cr^{2+} + 2NO_3^-$

$$M\,Cr^{2+} = \left(\frac{0.25 \text{ mol } Cr(NO_3)_2}{1\,L\,Cr(NO_3)_2 \text{ soln}}\right)\left(\frac{1 \text{ mol } Cr^{2+}}{1 \text{ mol } Cr(NO_3)_2}\right) = 0.25\,M\,Cr^{2+}$$

$$M\,NO_3^- = \left(\frac{0.25 \text{ mol } Cr(NO_3)_2}{1\,L\,Cr(NO_3)_2 \text{ soln}}\right)\left(\frac{2 \text{ mol } NO_3^-}{1 \text{ mol } Cr(NO_3)_2}\right) = 0.50\,M\,NO_3^-$$

(b) $CuSO_4 \rightarrow Cu^{2+} + SO_4^{2-}$

$$M\,Cu^{2+} = \left(\frac{0.10 \text{ mol } CuSO_4}{1\,L\,CuSO_4 \text{ soln}}\right)\left(\frac{1 \text{ mol } Cu^{2+}}{1 \text{ mol } CuSO_4}\right) = 0.10\,M\,Cu^{2+}$$

$$M\,SO_4^{2-} = \left(\frac{0.10 \text{ mol } CuSO_4}{1\,L\,CuSO_4 \text{ soln}}\right)\left(\frac{1 \text{ mol } SO_4^{2-}}{1 \text{ mol } CuSO_4}\right) = 0.10\,M\,SO_4^{2-}$$

(c) $Na_3PO_4 \rightarrow 3Na^+ + PO_4^{3-}$

$$M\,Na^+ = \left(\frac{0.16 \text{ mol } Na_3PO_4}{1\,L\,Na_3PO_4 \text{ soln}}\right)\left(\frac{3 \text{ mol } Na^+}{1 \text{ mol } Na_3PO_4}\right) = 0.48\,M\,Na^+$$

$$M\,PO_4^{3-} = \left(\frac{0.16 \text{ mol } Na_3PO_4}{1\,L\,Na_3PO_4 \text{ soln}}\right)\left(\frac{1 \text{ mol } PO_4^{3-}}{1 \text{ mol } Na_3PO_4}\right) = 0.16\,M\,PO_4^{3-}$$

(d) $Al_2(SO_4)_3 \rightarrow 2Al^{3+} + 3SO_4^{2-}$

$$M\,Al^{3+} = \left(\frac{0.075 \text{ mol } Al_2(SO_4)_3}{1\,L\,Al_2(SO_4)_3 \text{ soln}}\right)\left(\frac{2 \text{ mol } Al^{3+}}{1 \text{ mol } Al_2(SO_4)_3}\right) = 0.15\,M\,Al^{3+}$$

$$M\,SO_4^{2-} = \left(\frac{0.075 \text{ mol } Al_2(SO_4)_3}{1\,L\,Al_2(SO_4)_3 \text{ soln}}\right)\left(\frac{3 \text{ mol } SO_4^{2-}}{1 \text{ mol } Al_2(SO_4)_3}\right) = 0.23\,M\,SO_4^{2-}$$

4.61 $$M\,Na_3PO_4 = \left(\frac{0.21 \text{ mol } Na^+}{1\,L\,Na_3PO_4 \text{ soln}}\right)\left(\frac{1 \text{ mol } Na_3PO_4}{3 \text{ mol } Na^+}\right) = 0.070\,M\,Na_3PO_4$$

4.63

$$\text{\# g } Al_2(SO_4)_3 = (50.0 \text{ mL solution})\left(\frac{0.12 \text{ mol } Al^{3+}}{1000 \text{ mL solution}}\right)\left(\frac{1 \text{ mol } Al_2(SO_4)_3}{2 \text{ mol } Al^{3+}}\right)\left(\frac{342.14 \text{ g } Al_2(SO_4)_3}{1 \text{ mol } Al_2(SO_4)_3}\right)$$

$$= 1.0 \text{ g } Al_2(SO_4)_3$$

4.65

$$M\,Ba(OH)_2 = \frac{(20.78\,mL\,HCl\,soln)\left(\dfrac{1\,L\,HCl\,soln}{1000\,mL\,HCl\,soln}\right)\left(\dfrac{0.116\,mol\,HCl}{1\,L\,HCl\,soln}\right)\left(\dfrac{1\,mol\,Ba(OH)_2}{2\,mol\,HCl}\right)}{(21.34\,mL\,Ba(OH)_2\,soln)\left(\dfrac{1\,L\,Ba(OH)_2\,soln}{1000\,mL\,Ba(OH)_2\,soln}\right)}$$

$$= 0.0565\,M\,Ba(OH)_2$$

4.67 $Ag^+ + Cl^- \rightarrow AgCl(s)$

$$\#\,mL\,NiCl_2 = (20.0\,mL\,AgNO_3)\left(\frac{0.15\,mol\,AgNO_3}{1000\,mL\,AgNO_3}\right)\left(\frac{1\,mol\,Ag^+}{1\,mol\,AgNO_3}\right)\left(\frac{1\,mol\,Cl^-}{1\,mol\,Ag^+}\right)$$

$$\left(\frac{1\,mol\,NiCl_2}{2\,mol\,Cl^-}\right)\left(\frac{1000\,mL\,NiCl_2}{0.25\,mol\,NiCl_2}\right) = 6.0\,mL\,NiCl_2$$

$$\#\,g\,AgCl = (20.0\,mL\,AgNO_3)\left(\frac{0.15\,mol\,AgNO_3}{1000\,mL\,AgNO_3}\right)\left(\frac{1\,mol\,Ag^+}{1\,mol\,AgNO_3}\right)\left(\frac{1\,mol\,AgCl}{1\,mol\,Ag^+}\right)\left(\frac{143.32\,g\,Ag}{1\,mol\,AgCl}\right)$$

$$= 0.43\,g\,AgCl$$

4.69   First, determine the number of moles of H_3PO_4 that are to react:
0.0250 L x 0.250 mol/L = 6.25×10^{-3} mol H_3PO_4
Next, determine the number of moles of NaOH that are required by the balanced chemical equation: H_3PO_4 + $3NaOH \rightarrow Na_3PO_4 + 3H_2O$
6.25×10^{-3} mol H_3PO_4 x (3 mol NaOH/1 mol H_3PO_4) = 0.0188 mol NaOH

Last, calculate the volume of NaOH solution that will deliver this required number of moles of NaOH:

0.0188 mol NaOH × 1000 mL/0.100 mol = 188 mL

Alternatively, we can write:

0.0188 mol ÷ 0.100 mol/L = 0.188 L

4.71 $Ag^+ + Cl^- \rightarrow AgCl(s)$

$$\#\,mL\,AlCl_3 = (20.0\,AgC_2H_3O_2)\left(\frac{0.500\,mol\,AgC_2H_3O_2}{1000\,mL\,AgC_2H_3O_2}\right)\left(\frac{1\,mol\,Ag^+}{1\,mol\,AgC_2H_3O_2}\right)$$

$$\left(\frac{1\,mol\,Cl^-}{1\,mol\,Ag^+}\right)\left(\frac{1\,mol\,AlCl_3}{3\,mol\,Cl^-}\right)\left(\frac{1000\,mL\,AlCl_3}{0.250\,moles\,AlCl_3}\right) = 13.3\,mL\,AlCl_3$$

4.73 $Fe_2O_3 + 6HCl \rightarrow 2FeCl_3 + 3H_2O$
        0.0250 L HCl x 0.500 mol/L = 1.25 x 10^{-2} mol HCl

$$\# \text{ mol Fe}^{3+} = (1.25 \times 10^{-2} \text{ mol HCl}) \left(\frac{1 \text{ mol Fe}_2\text{O}_3}{6 \text{ mol HCl}} \right) \left(\frac{2 \text{ mol Fe}^{3+}}{1 \text{ mol Fe}_2\text{O}_3} \right)$$

$$= 4.17 \times 10^{-3} \text{ mol Fe}^{3+}$$

$$M \text{ Fe}^{3+} = \frac{4.17 \times 10^{-3} \text{ mol Fe}^{3+}}{0.0250 \text{ L soln}} = 0.167 \text{ M Fe}^{3+}$$

$$\# \text{ g Fe}_2\text{O}_3 = \left(4.17 \times 10^{-3} \text{ mol Fe}^{3+} \right) \left(\frac{1 \text{ mol Fe}_2\text{O}_3}{2 \text{ mol Fe}^{3+}} \right) \left(\frac{159.69 \text{ g Fe}_2\text{O}_3}{1 \text{ mol Fe}_2\text{O}_3} \right)$$

$$= 0.333 \text{ g Fe}_2\text{O}_3$$

Therefore, the mass of Fe_2O_3 that remains unreacted is:
$(4.00$ g $- 0.333$ g$) = 3.67$ g

4.75   The equation for the reaction indicates that the two materials react in equimolar amounts, i.e. the
        stoichiometry is 1 to 1:

$$AgNO_3(aq) + NaCl(aq) \rightarrow AgCl(s) + NaNO_3(aq)$$

(a)   Because this reaction is 1:1, we can see by inspection that the $AgNO_3$ is the limiting reagent. We
        know this because the concentration of the $AgNO_3$ is lower than the NaCl. Since we start with equal
        volumes, there are fewer moles of the $AgNO_3$.

$$\# \text{ mol AgCl} = (25.0 \text{ mL AgNO}_3 \text{ soln}) \left(\frac{0.320 \text{ mol AgNO}_3}{1000 \text{ mL AgNO}_3 \text{ soln}} \right)$$

$$\times \left(\frac{1 \text{ mol AgCl}}{1 \text{ mol AgNO}_3} \right)$$

$$= 8.00 \times 10^{-3} \text{ mol AgCl}$$

(b)   Assuming that AgCl is essentially insoluble, the concentration of silver ion can be said to be zero since
        all of the $AgNO_3$ reacted. The number of moles of chloride ion would be reduced by the precipitation
        of 8.00×10^{-3} mol AgCl, such that the final number of moles of chloride ion would be:
        0.0250 L $\times 0.440$ mol/L $- 8.00 \times 10^{-3}$ mol $= 3.0 \times 10^{-3}$ mol Cl^-

        The final concentration of Cl^- is, therefore:
            3.0×10^{-3} mol $\div 0.0500$ L $= 0.060$ M Cl^-

        All of the original number of moles of NO_3^- and of Na^+ would still be present in solution, and their
        concentrations would be:

        For NO_3^-:

$$\# \text{ M NO}_3^- = \frac{(25.0 \text{ mL AgNO}_3 \text{ soln}) \left(\frac{0.320 \text{ mol AgNO}_3}{1000 \text{ mL AgNO}_3 \text{ soln}} \right) \left(\frac{1 \text{ mol NO}_3^-}{1 \text{ mol AgNO}_3} \right)}{(50.0 \text{ mL soln}) \left(\frac{1 \text{ L soln}}{1000 \text{ mL soln}} \right)}$$

$$= 0.160 \text{ M NO}_3^-$$

For Na⁺:

$$\# \, M \, Na^+ = \frac{(25.0\,mL\;NaCl\;soln)\left(\dfrac{0.440\,mol\;NaCl}{1000\,mL\;NaCl\;soln}\right)\left(\dfrac{1\,mol\;Na^+}{1\,mol\;NaCl}\right)}{(50.0\,mL\;soln)\left(\dfrac{1\,L\;soln}{1000\,mL\;soln}\right)}$$

$$= 0.220\,M\,Na^+$$

4.77 $\# \, g \, Pb = (1.081\,g\;PbSO_4)\left(\dfrac{1\,mol\;BaSO_4}{303.27\,g\;BaSO_4}\right)\left(\dfrac{1\,mol\;Pb}{1\,mol\;PbSO_4}\right)\left(\dfrac{207.2\,g\;Pb}{1\,mol\;Pb}\right) = 0.7386\,g\,Pb$

The percentage of Pb in the sample can be calculated as

$$\% \, Pb = \left(\frac{mass\;of\;Pb}{mass\;of\;sample}\right) \times 100\% = \frac{0.7386\,g\;Pb}{1.526\,g\;PbSO_4} \times 100\% = 48.40\%\,Pb$$

4.79   First, calculate the number of moles HCl based on the titration according to the following equation:

$$NaOH(aq) + HCl(aq) \rightarrow NaCl(aq) + H_2O(\ell)$$

$$\# \, mol \, HCl = (23.25\,mL\;NaOH)\left(\frac{0.105\,mol\;NaOH}{1000\,mL\;NaOH}\right)\left(\frac{1\,mol\;HCl}{1\,mol\;NaOH}\right)$$

$$= 2.44 \times 10^{-3}\,mol\;HCl$$

Next, determine the concentration of the HCl solution:

$2.44 \times 10^{-3}\,mol \div 0.02145\,L = 0.114\,M\,HCl$

4.81   Since lactic acid is monoprotic, it reacts with sodium hydroxide on a one to one mole basis:

$$\# \, mol \, HC_3H_5O_3 = (17.25\,mL\;NaOH)\left(\frac{0.155\,mol\;NaOH}{1000\,mL\;NaOH}\right)\left(\frac{1\,mol\;HC_3H_5O_3}{1\,mol\;NaOH}\right)$$

$$= 2.67 \times 10^{-3}\,mol\;HC_3H_5O_3$$

4.83   (a) $\# \, mol \, BaSO_4 = (1.174\,g\;BaSO_4)\left(\dfrac{1\,mol\;BaSO_4}{233.39\,g\;BaSO_4}\right)$

$$= 5.030 \times 10^{-3}\,mol\;BaSO_4$$

(b)   As many moles as of $BaSO_4$, namely 5.030×10^{-3} mol $MgSO_4$. We know this because we can see that there is one mole of SO_4^{2-} in both the Ba and Mg compounds. Since both of these elements are in the same family, the reaction to produce the barium salt from the magnesium salt must be 1:1.

(c)   First determine the mass of $MgSO_4$ that was present:
5.030×10^{-3} mol $MgSO_4 \times 120.37$ g/mol $= 0.6055$ g $MgSO_4$
and subtract this from the total mass of the sample to find the mass of water in the original sample:
1.24 g $- 0.6055$ g $= 0.63$ g H_2O

(d)   We need to know the number of moles of water that are involved:
0.63 g $\div 18.0$ g/mol $= 3.5 \times 10^{-2}$ mol H_2O
and the relative mole amounts of water and $MgSO_4$:

For $MgSO_4$, 5.030×10^{-3} moles/5.030×10^{-3} moles = 1.000
For water, 3.5×10^{-2}/5.030×10^{-3} = 7.0
Hence the formula is $MgSO_4 7H_2O$

4.85 The HCl that is added will react with the Na_2CO_3. Therefore, the amount of Na_2CO_3 in the mixture may be determined from the amount of HCl needed to react with it. We determine this amount by measuring the amount of HCl added in excess. First determine the number of moles HCl added:

$$\text{\# moles HCl} = (50.0 \text{ mL HCl})\left(\frac{0.240 \text{ mol HCl}}{1000 \text{ mL HCl}}\right) = 1.20 \times 10^{-2} \text{ moles HCl}$$

Next determine how many moles of HCl remain after the reaction:

$$\text{\# moles HCl} = (22.90 \text{ mL NaOH})\left(\frac{0.100 \text{ mol NaOH}}{1000 \text{ mL NaOH}}\right)\left(\frac{1 \text{ mol HCl}}{1 \text{ mol NaOH}}\right)$$

$$= 2.29 \times 10^{-3} \text{ moles HCl}$$

Therefore, we know that;
1.20×10^{-2} moles HCl – 2.29×10^{-3} moles HCl = 9.71×10^{-3} moles HCl reacted with the Na_2CO_3 present in the mixture. The balanced equation for the reaction is:

$2 HCl + Na_2CO_3 \rightarrow 2NaCl + H_2O + CO_2$

$$\text{\# g Na2CO3} = (9.71 \times 10^{-3} \text{ mol HCl})\left(\frac{1 \text{ mol Na}_2CO_3}{2 \text{ mol HCl}}\right)\left(\frac{105.99 \text{ g Na}_2CO_3}{1 \text{ mol Na}_2CO_3}\right) = 0.515 \text{ g Na}_2CO_3$$

$\text{\# g NaCl} = 1.243 \text{ g} - 0.515 \text{g} = 0.728 \text{ g}$

$\% \text{NaCl} = \dfrac{0.728 \text{ g}}{1.243 \text{ g}} \times 100\% = 58.6\%$

Additional Exercises

4.87   (a)   strong electrolyte
       (b)   nonelectrolyte
       (c)   strong electrolyte
       (d)   non electrolyte
       (e)   weak electrolyte
       (f)   nonelectrolyte
       (g)   strong electrolyte
       (h)   weak electrolyte

4.89   (a)   molecular: $Na_2S(aq) + H_2SO_4(aq) \rightarrow H_2S(g) + Na_2SO_4(aq)$
            net ionic: $S^{2-}(aq) + 2H^+(aq) \rightarrow H_2S(g)$

       (b)   molecular: $LiHCO_3(aq) + HNO_3(aq) \rightarrow LiNO_3(aq) + H_2O(\ell) + CO_2(g)$
            net ionic: $H^+(aq) + HCO_3^-(aq) \rightarrow H_2O(\ell) + CO_2(g)$

       (c)   molecular: $(NH_4)_3PO_4(aq) + 3KOH(aq) \rightarrow$
                                     $K_3PO_4(aq) + 3NH_3(aq) + 3H_2O(\ell)$
            net ionic: $NH_4^+(aq) + OH^-(aq) \rightarrow NH_3(aq) + H_2O(\ell)$

       (d)   molecular: $K_2SO_3(aq) + HCl(aq) \rightarrow KHSO_3(aq) + KCl(aq)$
            net ionic: $SO_3^{2-}(aq) + H^+(aq) \rightarrow HSO_3^-(aq)$

(e)   molecular: $BaCO_3(s) + 2HBr(aq) \rightarrow BaBr_2(aq) + CO_2(g) + H_2O(\ell)$

net ionic: $BaCO_3(s) + 2H^+(aq) \rightarrow Ba^{2+}(aq) + CO_2(g) + H_2O(\ell)$

(f)   no reaction

4.91   0.02940 L KOH x 0.0300 mol/L = 8.82 $\times$ 10^{-4}mol KOH

8.82 $\times$ 10^{-4} mol KOH $\times$ (1 mol aspirin/1 mol KOH) = 8.82 $\times$ 10^{-4} mol aspirin

8.82 $\times$ 10^{-4} mol aspirin $\times$ 180.2 g/mol = 0.159 g aspirin

0.159 g aspirin/0.250 g total $\times$ 100 = 63.6 % aspirin

4.93   (a)   The mass of carbon in the original sample must be equal to the mass of carbon that is found in the CO_2.

$$\# \, g\,C = (22.17 \times 10^{-3}\, g\, CO_2)\left(\frac{1\, mole\, CO_2}{44.01 g\, CO_2}\right)\left(\frac{1\, mole\, C}{1\, mole\, CO_2}\right)\left(\frac{12.011 g\, C}{1\, mole\, C}\right)$$

$$= 6.051 \times 10^{-3}\, g\, C$$

Similarly, the entire mass of hydrogen that was present in the original sample ends up in the products as H_2O:

$$\# \, g\,H = (3.40 \times 10^{-3}\, g\, H_2O)\left(\frac{1\, mole\, H_2O}{18.02 g\, H_2O}\right)\left(\frac{2\, mole\, H}{1\, mole\, H_2O}\right)\left(\frac{1.008 g\, H}{1\, mole\, H}\right)$$

$$= 3.80 \times 10^{-4}\, g\, H$$

The mass of oxygen is determined by subtracting the mass due to C and H from the total mass:
10.46 mg total – (6.051 mg C + 0.380 mg H) = 4.03 mg O.

Convert these masses to mass percents:

$$\%\,C = \frac{6.051\, mg}{10.46\, mg} \times 100 = 57.85\,\%$$

$$\%\,H = \frac{0.380\, mg}{10.46\, mg} \times 100 = 3.63\,\%$$

$$\%\,O = \frac{4.03\, mg}{10.46\, mg} \times 100 = 38.5\,\%$$

(b)   Now, convert these masses to a number of moles:

$$\#\, moles\, C = \left(6.051 \times 10^{-3}\, g\, C\right)\left(\frac{1\, mole\, C}{12.011 g\, C}\right) = 5.038 \times 10^{-4}\, moles\, C$$

$$\#\, moles\, H = \left(3.80 \times 10^{-4}\, g\, H\right)\left(\frac{1\, mole\, H}{1.0079 g\, H}\right) = 3.77 \times 10^{-4}\, moles\, H$$

$$\#\, moles\, O = \left(4.03 \times 10^{-3}\, g\, O\right)\left(\frac{1\, mole\, O}{15.999 g\, O}\right) = 2.52 \times 10^{-4}\, moles\, O$$

The relative mole amounts are:

for C, $5.038 \times 10^{-4} \div 2.52 \times 10^{-4} = 2.00$
for H, $3.77 \times 10^{-4} \div 2.52 \times 10^{-4} = 1.50$
for O, $2.52 \times 10^{-4} \div 2.52 \times 10^{-4} = 1.00$

The empirical formula is $C_4H_3O_2$.

(c)   Since the empirical mass is 83 and the molecular mass is twice this amount (166), we conclude that the molecular formula must be $C_8H_6O_4$.

(d)   The number of moles of unknown acid used in the titration are:

$$\# \text{ mol acid} = (0.1680 \text{ g acid})\left(\frac{1 \text{ mol acid}}{166 \text{ g acid}}\right)$$

$$= 1.01 \times 10^{-3} \text{ mol acid}$$

The number of moles of base used in the titration are:

$$\# \text{ mol base} = (16.18 \text{ mL base})\left(\frac{0.1250 \text{ mol base}}{1000 \text{ mL base}}\right)$$

$$= 2.023 \times 10^{-3} \text{ mol base}$$

We therefore conclude that the unknown acid is diprotic since we used twice as many moles of base as moles of acid..

Practice Exercises

P5.1 $2Al(s) + 3Cl_2(g) \rightarrow 2AlCl_3(aq)$
Aluminum is oxidized and is, therefore, the reducing agent.
Chlorine is reduced and is, therefore, the oxidizing agent.

P5.2 (a) Ni +2; Cl –1
(b) Mg +2; Ti +4; O –2
(c) K +1; Cr +6; O –2
(d) H +1; P +5, O –2
(e) V +3; C 0; H +1; O –2

P5.3 +8/3

P5.4 $SnCl_3^- + 3Cl^- \rightarrow SnCl_6^{2-} + 2e^-$
$2HgCl_2 + 2e^- \rightarrow Hg_2Cl_2 + 2Cl^-$
$SnCl_3^- + 2HgCl_2 + Cl^- \rightarrow SnCl_6^{2-} + Hg_2Cl_2$

P5.5 $(Cu \rightarrow Cu^{2+} + 2e^-) \times 3$
$(NO_3^- + 4H^+ + 3e^- \rightarrow NO + 2H_2O) \times 2$
$3Cu + 2NO_3^- + 8H^+ \rightarrow 3Cu^{2+} + 2NO + 4H_2O$

P5.6 $(MnO_4^- + 4H^+ + 3e^- \rightarrow MnO_2 + 2H_2O) \times 2$
$(C_2O_4^{2-} + 2H_2O \rightarrow 2CO_3^{2-} + 4H^+ + 2e^-) \times 3$
$2MnO_4^- + 3C_2O_4^{2-} + 2H_2O \rightarrow 2MnO_2 + 6CO_3^{2-} + 4H^+$
Adding $4OH^-$ to both sides of the above equation we get:
$2MnO_4^- + 3C_2O_4^{2-} + 2H_2O + 4OH^- \rightarrow 2MnO_2 + 6CO_3^{2-} + 4H_2O$
which simplifies to give:
$2MnO_4^- + 3C_2O_4^{2-} + 4OH^- \rightarrow 2MnO_2 + 6CO_3^{2-} + 2H_2O$

P5.7 (a) molecular: $Mg(s) + 2HCl(aq) \rightarrow MgCl_2(aq) + H_2(g)$
ionic: $Mg(s) + 2H^+(aq) + 2Cl^-(aq) \rightarrow Mg^{2+}(aq) + 2Cl^-(aq) + H_2(g)$
net ionic: $Mg(s) + 2H^+(aq) \rightarrow Mg^{2+}(aq) + H_2(g)$

(b) molecular: $2Al(s) + 6HCl(aq) \rightarrow 2AlCl_3(aq) + 3H_2(g)$
ionic: $2Al(s) + 6H^+(aq) + 6Cl^-(aq) \rightarrow 2Al^{3+}(aq) + 6Cl^-(aq) + 3H_2(g)$
net ionic: $2Al(s) + 6H^+(aq) \rightarrow 2Al^{3+}(aq) + 3H_2(g)$

P5.8 (a) $2Al(s) + 3Cu^{2+}(aq) \rightarrow 2Al^{3+}(aq) + 3Cu(s)$
(b) N.R.

P5.9 $2C_4H_{10}(\ell) + 13O_2(g) \rightarrow 8CO_2(g) + 10H_2O(g)$

P5.10 $C_2H_5OH(\ell) + 3O_2(g) \rightarrow 2CO_2(g) + 3H_2O(g)$

P5.11 $4Fe(s) + 3O_2(g) \rightarrow 2Fe_2O_3(s)$

P5.12 $P_4(s) + 5O_2(g) \rightarrow P_4O_{10}(s)$

P5.13 First we need a balanced equation:
$Cl_2 + 2e^- \rightarrow 2Cl^-$
$S_2O_3^{2-} + 5H_2O \rightarrow 2SO_4^{2-} + 10H^+ + 8e^-$
$4Cl_2 + S_2O_3^{2-} + 5H_2O \rightarrow 8Cl^- + 2SO_4^{2-} + 10H^+$

$$\# \, g \, Na_2S_2O_3 = \left(4.25 \, g \, Cl_2\right)\left(\frac{1 \, mol \, Cl_2}{70.906 \, g \, Cl_2}\right)\left(\frac{1 \, mol \, Na_2S_2O_3}{4 \, mol \, Cl_2}\right)\left(\frac{158.132 \, g \, Na_2S_2O_3}{1 \, mol \, Na_2S_2O_3}\right)$$

$$= 2.37 \, g \, Na_2S_2O_3$$

P5.14   (a) $(Sn^{2+} \rightarrow Sn^{4+} + 2e^-) \times 5$

                   $(MnO_4^- + 8H^+ + 5e^- \rightarrow Mn^{2+} + 4H_2O) \times 2$

                   $5Sn^{2+} + 2MnO_4^- + 16H^+ \rightarrow 5Sn^{4+} + 2Mn^{2+} + 8H_2O$

        (b)

$$\# \, g \, Sn = \left(8.08 \, mL \, KMnO_4 \, soln\right)\left(\frac{0.0500 \, mol \, KMnO_4}{1000 \, mL \, KMnO_4}\right)\left(\frac{1 \, mol \, MnO_4^-}{1 \, mol \, KMnO_4}\right)$$

$$\times \left(\frac{5 \, mol \, Sn^{2+}}{2 \, mol \, MnO_4^-}\right)\left(\frac{1 \, mol \, Sn}{1 \, mol \, Sn^{2+}}\right)\left(\frac{118.71 \, g \, Sn}{1 \, mol \, Sn}\right)$$

$$= 0.120 \, g \, Sn$$

        (c) $\% \, Sn = \dfrac{0.120 \, g \, Sn}{0.300 \, g \, sample} \times 100 = 40.0 \, \% \, Sn$

        (d)

$$\# \, g \, Sn = \left(8.08 \, mL \, KMnO_4 \, soln\right)\left(\frac{0.0500 \, mol \, KMnO_4}{1000 \, mL \, KMnO_4}\right)\left(\frac{1 \, mol \, MnO_4^-}{1 \, mol \, KMnO_4}\right)$$

$$\times \left(\frac{5 \, mol \, Sn^{2+}}{2 \, mol \, MnO_4^-}\right)\left(\frac{1 \, mol \, SnO_2}{1 \, mol \, Sn^{2+}}\right)\left(\frac{150.71 \, g \, SnO_2}{1 \, mol \, SnO_2}\right)$$

$$= 0.152 \, g \, SnO_2$$

$$\% \, SnO_2 = \dfrac{0.152 \, g \, SnO_2}{0.300 \, g \, sample} \times 100 = 50.7 \, \% \, SnO_2$$

Review Problems

5.24   (a)   substance reduced (and oxidizing agent): HNO_3
                substance oxidized (and reducing agent): H_3AsO_3
        (b)   substance reduced (and oxidizing agent): $HOCl$
                substance oxidized (and reducing agent): NaI
        (c)   substance reduced (and oxidizing agent): $KMnO_4$
                substance oxidized (and reducing agent): $H_2C_2O_4$
        (d)   substance reduced (and oxidizing agent): H_2SO_4
                substance oxidized (and reducing agent): Al

5.26   Recall that the sum of the oxidation numbers must equal the charge on the molecule or ion.

        (a) -2
        (b) $+4$; we know that O is usually in a -2 oxidation state.
        (c)   The oxidation state of an element is always zero, by definition.
        (d) -3; hydrogen is usually in a $+1$ oxidation state.

5.28 The sum of the oxidation numbers should be zero:

(a)   Na: +1   (c)   Na: +1
      H: +1   S: +2.5
      P: +5   O: −2
      O: −2

(b)   Ba: +2   (d)   Cl: +3
      Mn: +6   F: −1
      O: −2

5.30 The sum of the oxidation numbers should be zero:

(a)   +2
(b)   +5
(c)   0
(d)   +4
(e)   −2

5.32 The sum of the oxidation numbers should be zero:

(a)   O: −2
      Na: +1
      Cl: +1
(b)   O: −2
      Na: +1
      Cl: +3
(c)   O: −2
      Na: +1
      Cl: +5
(d)   O: −2
      Na: +1
      Cl: +7

5.34 The sum of the oxidation numbers should be zero:

(a)   S: −2   (c)   O: −2
      Pb: +2   Sr: +2
                             I: +5

(b)   Cl: −1   (d)   S: −2
      Ti: +4   Cr: +3

5.36 (a) $BiO_3^- + 6H^+ + 2e^- \rightarrow Bi^{3+} + 3H_2O$ This is reduction of BiO_3^{2-}.
      (b) $Pb^{2+} + 2H_2O \rightarrow PbO_2 + 4H^+ + 2e^-$ This is oxidation of Pb^{2+}.

5.38 (a) $Fe + 2OH^- \rightarrow Fe(OH)_2 + 2e^-$ This is oxidation of Fe.
      (b) $2e^- + 2OH^- + SO_2Cl_2 \rightarrow SO_3^{2-} + 2Cl^- + H_2O$ This is reduction of SO_2Cl_2.

5.40 (a) $2S_2O_3^{2-} \rightarrow S_4O_6^{2-} + 2e^-$
            $OCl^- + 2H^+ + 2e^- \rightarrow Cl^- + H_2O$
            $OCl^- + 2S_2O_3^{2-} \rightarrow 2H^+ + S_4O_6^{2-} + Cl^- + H_2O$

      (b) $(NO_3^- + 2H^+ + e^- \rightarrow NO_2 + H_2O)^2$
            $Cu \rightarrow Cu^{2+} + 2e^-$
            $2NO_3^- + Cu + 4H^+ \rightarrow 2NO_2 + Cu^{2+} + 2H_2O$

(c) $IO_3^- + 6H^+ + 6e^- \rightarrow I^- + 3H_2O$

        $(H_2O + AsO_3^{3-} \rightarrow AsO_4^{3-} + 2H^+ + 2e^-) \times 3$

        $IO_3^- + 3AsO_3^{3-} + 6H^+ + 3H_2O \rightarrow I^- + 3AsO_4^{3-} + 3H_2O + 6H^+$

        which simplifies to give:

        $3AsO_3^{3-} + IO_3^- \rightarrow I^- + 3AsO_4^{3-}$

(d) $SO_4^{2-} + 4H^+ + 2e^- \rightarrow SO_2 + 2H_2O$

        $Zn \rightarrow Zn^{2+} + 2e^-$

        $Zn + SO_4^{2-} + 4H^+ \rightarrow Zn^{2+} + SO_2 + 2H_2O$

(e) $NO_3^- + 10H^+ + 8e^- \rightarrow NH_4^+ + 3H_2O$

        $(Zn \rightarrow Zn^{2+} + 2e^-) \times 4$

        $NO_3^- + 4Zn + 10H^+ \rightarrow 4Zn^{2+} + NH_4^+ + 3H_2O$

(f) $2Cr^{3+} + 7H_2O \rightarrow Cr_2O_7^{2-} + 14H^+ + 6e^-$

        $(BiO_3^- + 6H^+ + 2e^- \rightarrow Bi^{3+} + 3H_2O) \times 3$

        $2Cr^{3+} + 3BiO_3^- + 18H^+ + 7H_2O \rightarrow Cr_2O_7^{2-} + 14H^+ + 3Bi^{3+} + 9H_2O$

        which simplifies to give:

        $2Cr^{3+} + 3BiO_3^- + 4H^+ \rightarrow Cr_2O_7^{2-} + 3Bi^{3+} + 2H_2O$

(g) $I_2 + 6H_2O \rightarrow 2IO_3^- + 12H^+ + 10e^-$

        $(OCl^- + 2H^+ + 2e^- \rightarrow Cl^- + H_2O) \times 5$

        $I_2 + 5OCl^- + H_2O \rightarrow 2IO_3^- + 5Cl^- + 2H^+$

(h) $(Mn^{2+} + 4H_2O \rightarrow MnO_4^- + 8H^+ + 5e^-) \times 2$

        $(BiO_3^- + 6H^+ + 2e^- \rightarrow Bi^{3+} + 3H_2O) \times 5$

        $2Mn^{2+} + 5BiO_3^- + 30H^+ + 8H_2O \rightarrow 2MnO_4^- + 5Bi^{3+} + 16H^+ + 15H_2O$

        which simplifies to:

        $2Mn^{2+} + 5BiO_3^- + 14H^+ \rightarrow 2MnO_4^- + 5Bi^{3+} + 7H_2O$

(i) $(H_3AsO_3 + H_2O \rightarrow H_3AsO_4 + 2H^+ + 2e^-) \times 3$

        $Cr_2O_7^{2-} + 14H^+ + 6e^- \rightarrow 2Cr^{3+} + 7H_2O$

        $3H_3AsO_3 + Cr_2O_7^{2-} + 3H_2O + 14H^+ \rightarrow 3H_3AsO_4 + 2Cr^{3+} + 6H^+ + 7H_2O$

        which simplifies to give:

        $3H_3AsO_3 + Cr_2O_7^{2-} + 8H^+ \rightarrow 3H_3AsO_4 + 2Cr^{3+} + 4H_2O$

(j) $2I^- \rightarrow I_2 + 2e^-$

        $HSO_4^- + 3H^+ + 2e^- \rightarrow SO_2 + 2H_2O$

        $2I^- + HSO_4^- + 3H^+ \rightarrow I_2 + SO_2 + 2H_2O$

5.42   For redox reactions in basic solution, we proceed to balance the half reactions as if they were in acid solution, and then add enough OH^- to each side of the resulting equation in order to neutralize (titrate) all of the H^+. This gives a corresponding amount of water ($H^+ + OH^- \rightarrow H_2O$) on one side of the equation, and an excess of OH^- on the other side of the equation, as befits a reaction in basic solution.

      (a) $(CrO_4^{2-} + 4H^+ + 3e^- \rightarrow CrO_2^- + 2H_2O) \times 2$

            $(S^{2-} \rightarrow S + 2e^-) \times 3$

            $2CrO_4^{2-} + 3S^{2-} + 8H^+ \rightarrow 2CrO_2^- + 2S + 4H_2O$

            Adding $8OH^-$ to both sides of the above equation we obtain:

            $2CrO_4^{2-} + 3S^{2-} + 8H_2O \rightarrow 2CrO_2^- + 8OH^- + 3S + 4H_2O$

            which simplifies to:

            $2CrO_4^{2-} + 3S^{2-} + 4H_2O \rightarrow 2CrO_2^- + 3S + 8OH^-$

(b) $(C_2O_4^{2-} \rightarrow 2CO_2 + 2e^-)$ x 3

$(MnO_4^- + 4H^+ + 3e^- \rightarrow MnO_2 + 2H_2O)$ x 2

$3C_2O_4^{2-} + 2MnO_4^- + 8H^+ \rightarrow 6CO_2 + 2MnO_2 + 4H_2O$

Adding $8OH^-$ to both sides of the above equation we get:

$3C_2O_4^{2-} + 2MnO_4^- + 8H_2O \rightarrow 6CO_2 + 2MnO_2 + 4H_2O + 8OH^-$

which simplifies to give:

$3C_2O_4^{2-} + 2MnO_4^- + 4H_2O \rightarrow 6CO_2 + 2MnO_2 + 8OH^-$

(c) $(ClO_3^- + 6H^+ + 6e^- \rightarrow Cl^- + 3H_2O)$ x 4

$(N_2H_4 + 2H_2O \rightarrow 2NO + 8H^+ + 8e^-)$ x 3

$4ClO_3^- + 3N_2H_4 + 24H^+ + 6H_2O \rightarrow 4Cl^- + 6NO + 12H_2O + 24H^+$

which needs no OH^-, because it simplifies directly to:

$4ClO_3^- + 3N_2H_4 \rightarrow 4Cl^- + 6NO + 6H_2O$

(d) $NiO_2 + 2H^+ + 2e^- \rightarrow Ni(OH)_2$

$2Mn(OH)_2 \rightarrow Mn_2O_3 + H_2O + 2H^+ + 2e^-$

$NiO_2 + 2Mn(OH)_2 \rightarrow Ni(OH)_2 + Mn_2O_3 + H_2O$

(e) $(SO_3^{2-} + H_2O \rightarrow SO_4^{2-} + 2H^+ + 2e^-)$ x 3

$(MnO_4^- + 4H^+ + 3e^- \rightarrow MnO_2 + 2H_2O)$ x 2

$3SO_3^{2-} + 3H_2O + 8H^+ + 2MnO_4^- \rightarrow 3SO_4^{2-} + 6H^+ + 2MnO_2 + 4H_2O$

Adding $8OH^-$ to both sides of the equation we obtain:

$3SO_3^{2-} + 11H_2O + 2MnO_4^- \rightarrow 3SO_4^{2-} + 10H_2O + 2MnO_2 + 2OH^-$

which simplifies to:

$3SO_3^{2-} + 2MnO_4^- + H_2O \rightarrow 3SO_4^{2-} + 2MnO_2 + 2OH^-$

5.44 $(OCl^- + 2H^+ + 2e^- \rightarrow Cl^- + H_2O)$ x 4

$S_2O_3^{2-} + 5H_2O \rightarrow 2SO_4^{2-} + 10H^+ + 8e^-$

$4OCl^- + S_2O_3^{2-} + 5H_2O + 8H^+ \rightarrow 4Cl^- + 2SO_4^{2-} + 10H^+ + 4H_2O$

    which simplifies to:

$4OCl^- + S_2O_3^{2-} + H_2O \rightarrow 4Cl^- + 2SO_4^{2-} + 2H^+$

5.46   (a)   m: $Mn(s) + 2HCl(aq) \rightarrow MnCl_2(aq) + H_2(g)$

I: $Mn(s) + 2H^+(aq) + 2Cl^-(aq) \rightarrow Mn^{2+}(aq) + 2Cl^-(aq) + H_2(g)$

NI: $Mn(s) + 2H^+(aq) \rightarrow Mn^{2+}(aq) + H_2(g)$

(b)   m: $Cd(s) + 2HCl(aq) \rightarrow CdCl_2(aq) + H_2(g)$

I: $Cd(s) + 2H^+(aq) + 2Cl^-(aq) \rightarrow Cd^{2+}(aq) + Cl^-(aq) + H_2(g)$

NI: $Cd(s) + 2H^+(aq) \rightarrow Cd^{2+}(aq) + H_2(g)$

(c)   m: $Sn(s) + 2HCl(aq) \rightarrow SnCl_2(aq) + H_2(g)$

I: $Sn(s) + 2H^+(aq) + 2Cl^-(aq) \rightarrow Sn^{2+}(aq) + 2Cl^-(aq) + H_2(g)$

NI: $Sn(s) + 2H^+(aq) \rightarrow Sn^{2+}(aq) + H_2(g)$

(d)   m: $Ni(s) + 2HCl(aq) \rightarrow NiCl_2(aq) + H_2(g)$

I: $Ni(s) + 2H^+(aq) + 2Cl^-(aq) \rightarrow Ni^{2+}(aq) + 2Cl^-(aq) + H_2(g)$

NI: $Ni(s) + 2H^+(aq) \rightarrow Ni^{2+}(aq) + H_2(g)$

(e)   m: $2Cr(s) + 6HCl(aq) \rightarrow 2CrCl_3(aq) + 3H_2(g)$

I: $2Cr(s) + 6H^+(aq) + 6Cl^-(aq) \rightarrow 2Cr^{3+}(aq) + 6Cl^-(aq) + 3H_2(g)$

NI: $2Cr(s) + 6H^+(aq) \rightarrow 2Cr^{3+}(aq) + 3H_2(g)$

5.48   (a) $3Ag(s) + 4HNO_3(aq) \rightarrow 3AgNO_3(aq) + 2H_2O(\ell) + NO(g)$

         (b) $Ag(s) + 2HNO_3(aq) \rightarrow AgNO_3(aq) + H_2O(\ell) + NO_2(aq)$

5.50   In each case, the reaction should proceed to give the less reactive of the two metals, together with the ion of the more reactive of the two metals. The reactivity is taken from the reactivity series table 9.2.

         (a)   N.R.

         (b) $2Cr(s) + 3Pb^{2+}(aq) \rightarrow 2Cr^{3+}(aq) + 3Pb(s)$

         (c) $2Ag^{+}(aq) + Fe(s) \rightarrow 2Ag(s) + Fe^{2+}(aq)$

         (d) $3Ag(s) + Au^{3+}(aq) \rightarrow Au(s) + 3Ag^{+}(aq)$

5.52   (a) $2C_6H_6(\ell) + 15O_2(g) \rightarrow 12CO_2(g) + 6H_2O(g)$

         (b) $C_3H_8(g) + 5O_2(g) \rightarrow 3CO_2(g) + 4H_2O(g)$

         (c) $C_{21}H_{44}(s) + 32O_2(g) \rightarrow 21CO_2(g) + 22H_2O(g)$

5.54   (a) $2C_6H_6(\ell) + 9O_2(g) \rightarrow 12CO(g) + 6H_2O(g)$

                 $2C_3H_8(g) + 7O_2(g) \rightarrow 6CO(g) + 8H_2O(g)$

                 $2C_{21}H_{44}(s) + 43O_2(g) \rightarrow 42CO(g) + 44H_2O(g)$

         (b) $2C_6H_6(\ell) + 3O_2(g) \rightarrow 12C(s) + 6H_2O(g)$

                 $C_3H_8(g) + 2O_2(g) \rightarrow 3(s) + 4H_2O(g)$

                 $C_{21}H_{44}(s) + 11O_2(g) \rightarrow 21C(s) + 22H_2O(g)$

5.56 $2CH_3OH(\ell) + 3O_2(g) \rightarrow 2CO_2(g) + 4H_2O(g)$

5.58   (a) $[Mn^{2+}(aq) + 4H_2O(\ell) \rightarrow MnO_4^{-}(aq) + 8H^{+}(aq) + 5e^{-}]$ 2

               $[BiO_3^{-}(aq) + 6H^{+}(aq) + 2e^{-} \rightarrow Bi^{3+}(aq) + 3H_2O(\ell)]$ 5

               $2Mn^{2+}(aq) + 5BiO_3^{-}(aq) + 8H_2O(\ell) + 30H^{+}(aq) \rightarrow$

                               $2MnO_4^{-}(aq) + 5Bi^{3+}(aq) + 15H_2O(\ell) + 16H^{+}(aq)$

               which simplifies to give:

               $2Mn^{2+}(aq) + 5BiO_3^{-}(aq) + 14H^{+}(aq) \rightarrow 2MnO_4^{-}(aq) + 5Bi^{3+}(aq) + 7H_2O(\ell)$

         (b)

$$\# \text{ g NaBiO}_3 = \left(18.5 \text{ g Mn(NO}_3)_2\right)\left(\frac{1 \text{ mol Mn(NO}_3)_2}{179.0 \text{ g Mn(NO}_3)_2}\right)\left(\frac{1 \text{ mol Mn}^{2+}}{1 \text{ mol Mn(NO}_3)_2}\right)$$

$$\times \left(\frac{5 \text{ mol BiO}_3^{-}}{2 \text{ mol Mn}^{2+}}\right)\left(\frac{1 \text{ mol NaBiO}_3}{1 \text{ mol BiO}_3^{-}}\right)\left(\frac{280.0 \text{ g NaBiO}_3}{1 \text{ mol NaBiO}_3}\right)$$

$$= 72.3 \text{ g NaBiO}_3$$

5.60 $Cu + 2Ag^{+} \rightarrow Cu^{2+} + Ag$

$$\# \text{ g Cu} = (12.0 \text{ g Ag})\left(\frac{1 \text{ mol Ag}}{107.868 \text{ g Ag}}\right)\left(\frac{1 \text{ mol Cu}}{2 \text{ mol Ag}}\right)\left(\frac{63.54 \text{ g Cu}}{1 \text{ mol Cu}}\right) = 3.53 \text{ g Cu}$$

5.62   (a) $[MnO_4^{-} + 8H^{+} + 5e^{-} \rightarrow Mn^{2+} + 4H_2O]$ x 2

               $[Sn^{2+} \rightarrow Sn^{4+} + 2e^{-}]$ x 5

               $2MnO_4^{-} + 5Sn^{2+} + 16H^{+} \rightarrow 2Mn^{2+} + 5Sn^{4+} + 8H_2O$

$$\text{(b)} \quad \# \text{ mL KMnO}_4 = (40.0 \text{ mL SnCl}_2)\left(\frac{0.250 \text{ mol SnCl}_2}{1000 \text{ mL SnCl}_2}\right)\left(\frac{1 \text{ mol Sn}^{2+}}{1 \text{ mol SnCl}_2}\right)$$

$$\left(\frac{2 \text{ mol MnO}_4^-}{5 \text{ mol Sn}^{2+}}\right)\left(\frac{1 \text{ mol KMnO}_4}{1 \text{ mol MnO}_4^-}\right)\left(\frac{1000 \text{ mL KMnO}_4}{0.230 \text{ mol kMnO}_4}\right) = 17.4 \text{ mL}$$

5.64

$$\# \text{ mol Cu}^{2+} = (29.96 \text{ mL S}_2\text{O}_3^{2-})\left(\frac{0.02100 \text{ mol S}_2\text{O}_3^{2-}}{1000 \text{ mL S}_2\text{O}_3^{2-}}\right)$$

$$\times \left(\frac{1 \text{ mol I}_3^-}{2 \text{ mol S}_2\text{O}_3^{2-}}\right)\left(\frac{2 \text{ mol Cu}^{2+}}{1 \text{ mol I}_3^-}\right)$$

$$= 6.292 \times 10^{-4} \text{ mol Cu}^{2+}$$

$$\# \text{ g Cu} = (6.292 \times 10^{-4} \text{ mol Cu}) \ (63.546 \text{ g Cu/mol Cu})$$
$$= 3.998 \times 10^{-2} \text{ g Cu}$$

$$\% \text{ Cu} = (3.998 \times 10^{-2} \text{ g Cu}/0.4225 \text{ g sample}) \ 100 = 9.463 \ \%$$

(b)

$$\# \text{ g CuCO}_3 = (6.292 \times 10 - 4 \text{ mol CuCO}_3)\left(\frac{123.56 \text{ g CuCO}_3}{1 \text{ mol CuCO}_3}\right)$$

$$= 0.07774 \text{ g CuCO}_3$$

$$\% \text{ CuCO}_3 = \left(\frac{0.07774 \text{ g CuCO}_3}{0.4225 \text{ g sample}}\right) \times 100 = 18.40 \ \%$$

5.66   (a)

$$\# \text{ g H}_2\text{O}_2 = (17.60 \text{ mL KMnO}_4)\left(\frac{0.02000 \text{ mol KMnO}_4}{1000 \text{ mL KMnO}_4}\right)\left(\frac{1 \text{ mol MnO}_4^-}{1 \text{ mol KMnO}_4}\right)$$

$$\times \left(\frac{5 \text{ mol H}_2\text{O}_2}{2 \text{ mol MnO}_4^-}\right)\left(\frac{34.02 \text{ g H}_2\text{O}_2}{1 \text{ mol H}_2\text{O}_2}\right)$$

$$= 0.02994 \text{ g H}_2\text{O}_2$$

(b) 0.02994 g$/1.000$ g $100 = 2.994 \ \%$ H$_2$O$_2$

5.68   (a) $2\text{CrO}_4^{2-} + 3\text{SO}_3^{2-} + \text{H}_2\text{O} \rightarrow 2\text{CrO}_2^- + 3\text{SO}_4^{2-} + 2\text{OH}^-$

(b)

$$\# \text{ mol CrO}_4^{2-} = (3.18 \text{ g Na}_2\text{SO}_3)\left(\frac{1 \text{ mol Na}_2\text{SO}_3}{126.04 \text{ g Na}_2\text{SO}_3}\right)$$

$$\times \left(\frac{1 \text{ mol SO}_3^{2-}}{1 \text{ mol Na}_2\text{SO}_3}\right)\left(\frac{2 \text{ mol CrO}_4^{2-}}{3 \text{ mol SO}_3^{2-}}\right)$$

$$= 1.68 \times 10^{-2} \text{ mol CrO}_4^{2-}$$

Since there is one mole of Cr in each mole of CrO_4^{2-}, then the above number of moles of CrO_4^{2-} is also equal to the number of moles of Cr that were present:

0.0168 mol Cr 52.00 g/mol = 0.875 g Cr in the original alloy.

(c)   (0.875 g/3.450 g) 100 = 25.4 % Cr

5.70   (a)

$$\# \, mol \, C_2O_4^{2-} = (21.62 \, mL \, KMnO_4)\left(\frac{0.1000 \, mol \, KMnO_4}{1000 \, mL \, KMnO_4}\right)\left(\frac{5 \, mol \, C_2O_4^{2-}}{2 \, mol \, KMnO_4}\right)$$

$$= 5.405 \times 10^{-3} \, mol \, C_2O_4^{2-}$$

(b)   The stoichiometry for calcium is as follows:
mol $C_2O_4^{2-}$ = mol Ca^{2+} = mol $CaCl_2$
Thus the number of grams of $CaCl_2$ is given simply by:
5.405×10^{-3} mol $CaCl_2$ x 110.98 g/mol = 0.5999 g $CaCl_2$

(c)   (0.5999 g/2.463 g) x 100 = 24.35 % $CaCl_2$

Additional Exercises

5.73   The first reaction demonstrates that Al is more readily oxidized than Cu. The second reaction demonstrates that Al is more readily oxidized than Fe. Reaction 3 demonstrates that Fe is more readily oxidized than Pb. Reaction 4 demonstrates that Fe is more readily oxidized than Cu. The fifth reaction demonstrates that Al is more readily oxidized than Pb. The last reaction demonstrates that Pb is more readily oxidized than Cu.

Altogether, the above facts constitute the following trend of increasing ease of oxidation:
Cu < Pb < Fe < Al

5.75   Any metal that is lower than hydrogen in the activity series shown in Table 5.2 of the text will react with H^+:
(c) zinc and (d) magnesium.

5.77   In each case, the reaction should proceed to give the less reactive of the two metals, together with the ion of the more reactive of the two metals. The reactivity is taken from the reactivity series.
(a) $Zn(s) + Sn^{2+}(aq) \rightarrow Zn^{2+}(aq) + Sn(s)$
(b) $2Cr(s) + 6H^+(aq) \rightarrow 2Cr^{3+}(aq) + 3H_2(g)$
(c)   N.R.
(d) $Mn(s) + Pb^{2+}(aq) \rightarrow Mn^{2+}(aq) + Pb(s)$
(e) $Zn(s) + Co^{2+}(aq) \rightarrow Zn^{2+}(aq) + Co(s)$

5.79   (a) $2Zn(s) + O_2(g) \rightarrow 2ZnO(s)$
(b) $4Al(s) + 3O_2(g) \rightarrow 2Al_2O_3(s)$
(c) $2Mg(s) + O_2(g) \rightarrow 2MgO(s)$
(d) $2Fe(s) + O_2(g) \rightarrow 2FeO(s)$
alternatively we have: $4Fe(s) + 3O_2(g) \rightarrow 2Fe_2O_3(s)$
(e) $2Ca(s) + O_2(g) \rightarrow 2CaO(s)$

5.81
$$\# \, g \, PbO_2 = (15.0 \, g \, Cl_2)\left(\frac{1 \, mol \, Cl_2}{70.91 \, g \, Cl_2}\right)\left(\frac{1 \, mol \, PbO_2}{1 \, mol \, Cl_2}\right)\left(\frac{239.2 \, g \, PbO_2}{1 \, mol \, PbO_2}\right)$$

$$= 50.6 \, g \, PbO_2$$

5.83 $Cu + 2Ag^+ \rightarrow 2Ag + Cu^{2+}$

The number of moles of Ag^+ available for the reaction is
$0.125\ M \times 0.255\ L = 0.0319\ mol\ Ag^+$

Since the stoichiometry is 2/1, the number of moles of Cu^{2+} ion that are consumed is $0.0319 \div 2 = 0.0159\ mol$. The mass of copper consumed is $0.0159\ mol \times 63.546\ g/mol = 1.01\ g$. The amount of unreacted copper is thus: $12.340\ g - 1.01\ g = 11.33\ g\ Cu$. The mass of Ag that is formed is: $0.0319\ mol \times 108\ g/mol = 3.45\ g$ Ag. The final mass of the bar is: $11.33\ g + 3.45\ g = 14.78\ g$.

5.85   The balanced equation for the oxidation-reduction reaction is:
$3H_2C_2O_4 + Cr_2O_7^{2-} + 8H^+ \rightarrow 6CO_2 + 2Cr^{3+} + 7H_2O$

$$\# \text{ moles } H_2C_2O_4 = \left(6.25\text{ mL } K_2Cr_2O_7\right)\left(\frac{0.200 \text{ moles } K_2Cr_2O_7}{1000 \text{ mL } K_2Cr_2O_7}\right)\left(\frac{3 \text{ moles } H_2C_2O_4}{1 \text{ mole } K_2Cr_2O_7}\right)$$

$$= 3.75 \times 10^{-3} \text{ moles } H_2C_2O_4$$

So, if we titrate the same oxalic acid solution using NaOH we will need:

$$\# \text{ ml NaOH} = \left(3.75 \times 10^{-3} \text{ moles } H_2C_2O_4\right)\left(\frac{2 \text{ moles NaOH}}{1 \text{ mole } H_2C_2O_4}\right)\left(\frac{1000 \text{ mL NaOH}}{0.450 \text{ moles NaOH}}\right)$$

$$= 16.7 \text{ ml NaOH}$$

Practice Exercises

P6.1   The amount of heat transferred into the water is:

$$\# J = (250 \text{ g H}_2\text{O})(4.184 \text{ } J/_{g \cdot {}^{\circ}C})(30.0 \text{ }^{\circ}C - 25.0 \text{ }^{\circ}C) = 5200 \text{ J}$$

$$\# kJ = (5200 \text{ J})\left(\frac{1 \text{ kJ}}{1000 \text{ J}}\right) = 5.2 \text{ kJ}$$

$$\# cal = (5200 \text{ J})\left(\frac{1 \text{ cal}}{4.184 \text{ J}}\right) = 1200 \text{ cal}$$

$$\# kcal = (1200 \text{ cal})\left(\frac{1 \text{ kcal}}{1000 \text{ cal}}\right) = 1.2 \text{ kcal}$$

P6.2   q = specific heat x mass x temperature change
        = 4.184 J/g °C x (175 g + 4.90 g) x (14.9 °C – 10.0 °C)
        = 3.7 X 10³ J = 3.7 kJ of heat released by the process.

This should then be converted to a value representing kJ per mole of reactant, remembering that the sign of ΔH is to be negative, since the process releases heat energy to surroundings. The number of moles of sulfuric acid is:

$$\# \text{ moles H}_2\text{SO}_4 = (4.90 \text{ g H}_2\text{SO}_4)\left(\frac{1 \text{ mole H}_2\text{SO}_4}{98.08 \text{ g H}_2\text{SO}_4}\right) = 5.00 \text{ X } 10^{-2} \text{ moles H}_2\text{SO}_4$$

and the enthalpy change in kJ/mole is given by:

3.7 kJ ÷ 0.0500 moles = 74 kJ/mole

P6.3   We can proceed by multiplying both the equation and the thermochemical value of Example 6.3 by the fraction 2.5/2:

2.5 H₂(g) + 2.5/2 O₂(g) → 2.5 H₂O(g),   ΔH = (–517.8 kJ) x 2.5/2
2.5 H₂(g) + 1.25 O₂(g) → 2.5 H₂O(g),   ΔH = –647.3 kJ

P6.4

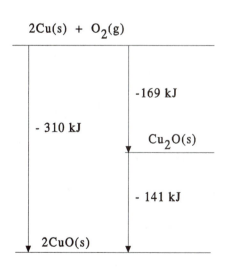

P6.5   This problem requires that we add the reverse of the second equation (remembering to change the sign of the associated ΔH value) to the first equation:

$$C_2H_4(g) + 3O_2(g) \rightarrow 2CO_2(g) + 2H_2O(\ell), \qquad\qquad \Delta H° = -1411.1 \text{ kJ}$$
$$2CO_2(g) + 3H_2O(\ell) \rightarrow C_2H_5OH(\ell) + 3O_2(g), \qquad\qquad \Delta H° = +1367.1 \text{ kJ}$$

which gives the following net equation and value for ΔH°:

$$C_2H_4(g) + H_2O(\ell) \rightarrow C_2H_5OH(\ell), \qquad\qquad \Delta H° = -44.0 \text{ kJ}$$

P6.6 $Na(s) + 1/2H_2(g) + C(s) + 3/2O_2(g) \rightarrow NaHCO_3(s), \quad \Delta H_f° = -947.7 \text{ kJ/mol}$

P6.7   (a) $\Delta H° = $ sum $\Delta H_f°$[products] − sum $\Delta H_f°$[reactants]
         $= 2\Delta H_f°[NO_2(g)] - \{2\Delta H_f°[NO(g)] + \Delta H_f°[O_2(g)]\}$
         $= 2$ mol x 33.8 kJ/mol − [2 mol x 90.37 kJ/mol + 1 mol x 0 kJ/mol]
         $= -113.1$ kJ

        (b) $\Delta H° = \{\Delta H_f°[H_2O(\ell)] + \Delta H_f°[NaCl(s)]\} - \{\Delta H_f°[NaOH(s)] + \Delta H_f°[HCl(g)]\}$
         $= [(-285.9 \text{ kJ/mol}) + (-411.0 \text{ kJ/mol})]$
                $- [(-426.8 \text{ kJ/mol}) + (-92.30 \text{ kJ/mol})]$
         $= -177.8$ kJ

Review Problems

6.34 $KE = \frac{1}{2} mv^2$

$$= 1/2(2150 \text{ kg})(80\frac{\text{km}}{\text{hr}})^2\left(\frac{1\,\text{hr}}{3600\text{s}}\right)^2\left(\frac{1000\text{m}}{1\,\text{km}}\right)^2$$

$$= (5.31 \times 10^5 \text{ kgm}^2/s^2)\left(\frac{1\,\text{kJ}}{1000\,\text{kgm}^2/s^2}\right)$$

$$= 531 \text{ kJ}$$

6.36   (a) $\#\,k\,cal = (457\,\text{kJ})\left(\frac{1\,k\,cal}{4.184\,\text{kJ}}\right) = 109\,k\,cal$

        (b) $\#\,\text{kJ} = (127\,k\,cal)\left(\frac{4.184\,\text{kJ}}{1\,k\,cal}\right) = 531\,\text{kJ}$

6.38 $\Delta E = q + w = 28J - 45J = -17 \text{ J}$

6.40 $\#\,\text{kJ} = \left(175\,\text{g H}_2\text{O}\right)\left(4.184\frac{J}{g\,°C}\right)\left(25.0\,°C - 15.0\,°C\right)\left(\frac{1\,\text{kJ}}{1000\,\text{J}}\right) = 7.32 \text{ kJ}$

6.42 $\#\,J = 0.4498\,J\,g^{-1}\,°C^{-1}$ x 15.0 g x 20.0 °C = 135 J

6.44   (a) $\#\,J = 4.184\,J\,g^{-1}\,°C^{-1}$ x 100 g x 4.0 °C = 1.67 X 10³ J
        (b)   1.67 X 10³ J
        (c)   1.67 X 10³ J/(100 − 28.0) °C = 23.2 J °C⁻¹
        (d)   23.2 J °C⁻¹ ÷ 50.0 g = 0.464 J g⁻¹ °C⁻¹

6.46   (a)   Since fat tissue is 85% fat, there are only 0.85 lbs of fat lost for every pound of tissue lost in the weight reduction program. The water, which is a part of the fat tissue, is not lost. Therefore;

$$\# \text{ kcal} = (0.85 \text{ lb fat})\left(\frac{456 \text{ g}}{1 \text{ lb}}\right)\left(\frac{9.0 \text{ kcal}}{1 \text{ g fat}}\right) = 3.5 \times 10^3 \text{ kcal}$$

        (b)   In part (a) we determined that we would need to expend 3500 kcal in order to burn off 1 lb of fat tissue.

$$\# \text{ miles} = (1.0 \text{ lb fat tissue})\left(\frac{3.5 \times 10^3 \text{ kcal}}{1.0 \text{ lb fat tissue}}\right)\left(\frac{1 \text{ hr}}{5.0 \times 10^2 \text{ kcal}}\right)\left(\frac{8.0 \text{ miles}}{\text{hr}}\right)$$

$$= 56 \text{ miles}$$

6.48 $$\frac{\# \text{ J}}{\text{mol }^\circ\text{C}} = \left(\frac{0.4498 \text{ J}}{\text{g }^\circ\text{C}}\right)\left(\frac{55.847 \text{ g Fe}}{1 \text{ mol Fe}}\right) = 25.12 \text{ J}\big/_{\text{mol }^\circ\text{C}}$$

6.50 $4.18 \text{ J g}^{-1}{}^\circ\text{C}^{-1} \times 4.54 \times 10^3 \text{ g} \times (58.65 - 60.25)\,^\circ\text{C} = -3.04 \times 10^4 \text{ J} = -30.4 \text{ kJ}$

6.52   Keep in mind that the total mass must be considered in this calculation, and that both liquids, once mixed, undergo the same temperature increase:
heat $= (4.18 \text{ J/g }^\circ\text{C}) \times (55.0 \text{ g} + 55.0 \text{ g}) \times (31.8\,^\circ\text{C} - 23.5\,^\circ\text{C})$
$\quad\quad = 3.8 \times 10^3 \text{ J}$ of heat energy released

Next determine the number of moles of reactant involved in the reaction:
0.0550 L x 1.3 mol/L = 0.072 mol of acid and of base.

Thus the enthalpy change is: $$\frac{\# \text{ kJ}}{\text{mol}} = \frac{(3.8 \times 10^3 \text{ J})\left(\dfrac{1 \text{ kJ}}{1000 \text{ J}}\right)}{(0.072 \text{ mol})} = 53 \text{ kJ}\big/_{\text{mol}}$$

6.54   (a) $\# \text{ J} = (97.1 \text{ kJ}/^\circ\text{C})(27.282\,^\circ\text{C} - 25.000\,^\circ\text{C}) = 222 \text{ kJ} = 2.22 \times 10^3 \text{ J}$
        (b) $\Delta H^\circ = -222$ kJ/mol

6.56   (a)   Multiply the given equation by the fraction 2/3.
                $2\text{CO(g)} + \text{O}_2\text{(g)} \rightarrow 2\text{CO}_2\text{(g)}, \quad \Delta H^\circ = -566 \text{ kJ}$

        (b)   To determine ΔH for 1 mol, simply multiply the original ΔH by 1/3; -283 kJ/mol.

6.58 $4\text{Al(s)} + 2\text{Fe}_2\text{O}_3\text{(s)} \rightarrow 2\text{Al}_2\text{O}_3\text{(s)} + 4\text{Fe(s)} \quad\quad \Delta H^\circ = -1708 \text{ kJ}$

6.60 $$\# \text{ kJ} = (6.54 \text{ g Mg})\left(\frac{-1203 \text{ kJ}}{2 \text{ mol Mg}}\right)\left(\frac{1 \text{ mol Mg}}{24.305 \text{ g Mg}}\right) = -162 \text{ kJ}, 162 \text{ kJ of heat are evolved}$$

6.62

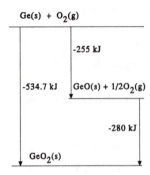

The enthalpy change for the reaction $GeO(s) + 1/2O_2(g) \rightarrow GeO_2(s)$ is –280 kJ as seen in the figure above.

6.64   Since NO_2 does not appear in the desired overall reaction, the two steps are to be manipulated in such a manner so as to remove it by cancellation. Add the second equation to the inverse of the first, remembering to change the sign of the first equation, since it is to be reversed:

$2NO_2(g) \rightarrow N_2O_4(g)$, $\Delta H° = -57.93$ kJ
$2NO(g) + O_2(g) \rightarrow 2NO_2(g)$, $\Delta H° = -113.14$ kJ

Adding, we have:

$2NO(g) + O_2(g) \rightarrow N_2O_4(g)$, $\Delta H° = -171.07$ kJ

6.66   If we label the four known thermochemical equations consecutively, 1, 2, 3, and 4, then the sum is made in the following way: Divide equation #3 by two, and reverse all of the other equations (#1, #2, and #4), while also dividing each by two:

$Na_2O(s) + HCl(g) \rightarrow H_2O(\ell) + NaCl(s)$, $\Delta H° = -253.66$ kJ
$NaNO_2(s) \rightarrow Na_2O(s) + NO_2(g) + NO(g)$, $\Delta H° = 213.57$ kJ
$NO(g) + NO_2(g) \rightarrow N_2O(g) + O_2(g)$, $\Delta H° = -21.34$ kJ
$H_2O(\ell) + O_2(g) + N_2O(g) \rightarrow HNO_2(\ell)$, $\Delta H° = -17.18$ kJ

Adding gives:

$HCl(g) + NaNO_2(s) \rightarrow NaCl(s) + HNO_2(\ell)$, $\Delta H° = -78.61$ kJ

6.68   Reverse the second equation, and then divide each by two before adding:

$CO(g) + O_2(g) \rightarrow CO_2(g)$, $\Delta H° = -283.0$ kJ
$CuO(s) \rightarrow Cu(s) + O_2(g)$, $\Delta H° = 155.2$ kJ

$CuO(s) + CO(g) \rightarrow Cu(s) + CO_2(g)$, $\Delta H° = -127.8$ kJ

6.70   Multiply all of the equations by ½ and add them together.

$1/2CaO(s) + 1/2Cl_2(g) \rightarrow 1/2CaOCl_2(s)$ $\Delta H° = ½(-110.9kJ)$

$1/2H_2O(\ell) + 1/2CaOCl_2(s) + NaBr(s) \rightarrow NaCl(s) + 1/2Ca(OH)_2(s) + 1/2Br_2(\ell)$
$\Delta H° = ½(-60.2$ kJ$)$

$1/2Ca(OH)_2(s) \rightarrow 1/2CaO(s) + 1/2H_2O(\ell)$ $\Delta H° = \frac{1}{2}(+65.1 \text{ kJ})$

$1/2Cl_2(g) + NaBr(s) \rightarrow NaCl(s) + 1/2Br_2(\ell)$ $\Delta H° = \frac{1}{2}(-106 \text{ kJ})$

6.72   We need to eliminate the NO_2 from the two equations. To do this, multiply the first reaction by 3 and the second reaction by two and add them together.

$12NH3(g) + 21O2(g) \rightarrow 12NO2(g) + 18H2O(g)$ $\Delta H° = 3(-1132 \text{ kJ})$

$12NO2(g) + 16NH3(g) \rightarrow 14N2(g) + 24H2O(g)$ $\Delta H° = 2(-2740 \text{ kJ})$

$28NH3(g) + 21O2(g) \rightarrow 14N2(g) + 42O2(g)$ $\Delta H° = -8876 \text{ kJ}$

Now divide this equation by 7 to get
$4NH3(g) + 3O2(g) \rightarrow 2N2(g) + 6O2(g)$ $\Delta H° = 1/7(-8876 \text{ kJ}) = -1268 \text{ kJ}$

6.74   The equation we want is:

$3Mg(s) + N_2(g) + 3O_2(g) \rightarrow Mg(NO_3)_2(s)$

Reverse all three reactions AND multiply the third equation by three

$Mg_3N_2(s) + 6MgO(s) \rightarrow 8Mg(s) + Mg(NO_3)_2(s)$ $\Delta H° = +3884 \text{ kJ}$

$3Mg(s) + N_2(g) \rightarrow Mg_3N_2(s)$ $\Delta H° = -463 \text{ kJ}$

$6Mg(s) + 3O_2(g) \rightarrow Mg(NO_3)_2(s)$ $\Delta H° = 3(-1203 \text{ kJ})$

$Mg(s) + N_2(g) + 3O_2(g) \rightarrow Mg(NO_3)_2(s)$ $\Delta H° = -188 \text{ kJ}$

6.76   Only (b) should be labeled with $\Delta H_f°$.

6.78   (a) $2C(graphite) + 2H_2(g) + O_2(g) \rightarrow HC_2H_3O_2(\ell)$, $\Delta H_f° = -487.0 \text{ kJ}$
         (b) $Na(s) + 1/2H_2(g) + C(graphite) + 3/2O_2(g) \rightarrow NaHCO_3(s)$ $\Delta H_f° = -947.7 \text{ kJ}$

         (c) $Ca(s) + 1/8S_8(s) + 3O_2(g) + 2H_2(g) \rightarrow CaSO_4\cdot2H_2O(s)$ $\Delta H_f° = -2021.1 \text{ kJ}$

6.80   (a) $\Delta H° = \Delta H_f°[O_2(g)] + 2\Delta H_f°[H_2O(\ell)] - 2\Delta H_f°[H_2O_2(\ell)]$

                $\Delta H° = 0 \text{ kJ/mol} + 2 \text{ mol} \times (-285.9 \text{ kJ/mol}) - 2 \times (-187.6 \text{ kJ/mol})$
                $= -196.6 \text{ kJ}$

         (b) $\Delta H° = \Delta H_f°[H_2O(\ell)] + \Delta H_f°[NaCl(s)] - \Delta H_f°[HCl(g)] - \Delta H_f°[NaOH(s)]$
                $= 1 \text{ mol} \times (-285.9 \text{ kJ/mol}) + 1 \text{ mol} \times (-411.0 \text{ kJ/mol})$
                $\quad - 1 \text{ mol} \times (-92.30 \text{ kJ/mol}) - 1 \text{ mol} \times (-426.8 \text{ kJ/mol})$
                $= -177.8 \text{ kJ}$

6.82   (a) $\frac{1}{2}H_2(g) + \frac{1}{2}Cl_2(g) \rightarrow HCl(g)$, $\Delta H_f° = -92.30 \text{ kJ/mol}$
         (b) $\frac{1}{2}N_2(g) + 2H_2(g) + \frac{1}{2}Cl_2(g) \rightarrow NH_4Cl(s)$, $\Delta H_f° = -315.4 \text{ kJ/mol}$

6.84 $C_{12}H_{22}O_{11}(s) + 12O_2(g) \rightarrow 12CO_2(g) + 11H_2O(\ell)$ $\Delta H° = -5.65 \times 10^3 \text{ kJ/mol}$

$\Delta H° = \Sigma\Delta H_f°(products) - \Sigma\Delta H_f°(reactants)$
$= [12\,\Delta H_f°(CO_2(g)) + 11\,\Delta H_f°(H_2O(\ell))] - [\Delta H_f°(C_{12}H_{22}O_{11}(s)) + 12\Delta H_f°(O_2(g))]$

Rearranging and realizing the $\Delta H_f°O_2(g) = 0$ we get

$\Delta H_f°(C_{12}H_{22}O_{11}(s)) = 12\Delta H_f°(CO_2(g)) + 11\Delta H_f°(H2O(\ell)) - \Delta H°$

$= 12(-393kJ) + 11(-285.9\ kJ) - (-5.65 \times 10^3\ kJ) = -2.22 \times 10^3\ kJ$

Additional Exercises

6.87 Heat lost = – Heat gained

$\text{Heat lost} = (1.000kg)\left(\dfrac{1000\ g}{1\ kg}\right)(0.4498\ J/g°C)(100.0°C - x)$

$\text{Heat gained} = (2.000\ kg)\left(\dfrac{1000\ g}{1\ kg}\right)(4.184\ J/g°C)(25.00\ °C - x)$

$44.98\ J/g°C(100.0°C - x) = -(836.8\ J/g°C)(25.00°C - x)$

$25418\ J = (881.8\ J/g°C)(x)$

$x = 25418\ J/881.8\ J/g°C = 28.83°C$

6.89 $\Delta H° = 3\Delta H_f°[CO_2(g)] + 4\Delta H_f°[Fe(s)] - 2\Delta H_f°[Fe_2O_3(s)] - 3\Delta H_f°[C(s)]$

$= 3\ mol \times (-393.5\ kJ/mol) + 4\ mol \times (0.0\ kJ/mol)$

$- 2\ mol \times (-822.2\ kJ/mol) - 3\ mol \times (0.0\ kJ/mol)$

$= 463.9\ kJ$

This reaction is endothermic.

6.90 Multiply the first reaction by ½:

$!/2Fe_2O_3(s) + 3/2CO(g) \rightarrow Fe(s) + 3/2CO_2(g)$　　　　　　$\Delta H° = ½(-28\ kJ)$

Reverse the second reaction AND multiply by 1/6:

$1/3Fe_3O_4(s) + 1/6CO_2(g) \rightarrow 1/3\ Fe_3O_4(s) + 1/6CO(g)$　$\Delta H° = 1/6(+59\ kJ)$

Reverse the third reaction AND multiply by 1/3:

$FeO(s) + 1/3CO_2(g) \rightarrow 1/3Fe_3O_4(s) + 1/3CO(g)$　　　　$\Delta H° = 1/3(-38\ kJ)$

Now add the equations together:

$FeO(s) + CO(g) \rightarrow Fe(s) + CO_2(g)$　　　　　　　　　　$\Delta H° = -16.8\ kJ$

6.91 $\Delta H° = [\Delta H_f°(CO_2(g)) + \Delta H_f°(Fe(s))] - [\Delta H_f°(FeO(s)) - \Delta H_f°(CO(g))]$

Rearranging, and remembering that $\Delta H_f°Fe(s) = 0$

$\Delta H_f°(FeO(s)) = \Delta H_f°(CO(g)) - \Delta H_f°(CO_2(g)) - \Delta H°$

$= -393.5\ kJ - (-110.5\ kJ) - (-16.8\ kJ) = -266.2\ kJ$

6.94 The desired net equation is obtained by adding together the reverse of the two thermochemical equations:

$Zn(NO_3)_2(aq) + Cu(s) \rightarrow Cu(NO_3)_2(aq) + Zn(s),$　　$\Delta H° = 258\ kJ$

$Cu(NO_3)_2(aq) + 2Ag(s) \rightarrow 2AgNO_3(aq) + Cu(s),$　　$\Delta H° = 106\ kJ$

$2Ag(s) + Zn(NO_3)_2(aq) \rightarrow Zn(s) + 2AgNO_3(aq),$　　$\Delta H° = 364\ kJ$

Since this $\Delta H°$ has a positive value, it is the reverse reaction that occurs spontaneously.

6.96   The equation may be written as: $1/2\ H_2(g) + 1/2\ Br_2(\ell) \rightarrow HBr(g)$; $\Delta H_f^\circ = -36$ kJ

To obtain ΔH, combine the equations in the following manner:

$Br_2(aq) + 2KCl(aq)$	$\rightarrow$	$Cl_2(g) + 2KBr(aq)$	$\Delta H^\circ = 96.2$ kJ
$H_2(g) + Cl_2(g)$	$\rightarrow$	$2HCl(g)$	$\Delta H^\circ = -184$ kJ
$2HCl(aq) + 2KOH(aq)$	$\rightarrow$	$2KCl(aq) + 2H_2O(\ell)$	$\Delta H^\circ = -115$ kJ
$2KBr(aq) + 2H_2O(\ell)$	$\rightarrow$	$2HBr(aq) + 2KOH(aq)$	$\Delta H^\circ = 115$ kJ
$2HCl(g)$	$\rightarrow$	$2HCl(aq)$	$\Delta H^\circ = -154$ kJ
$2HBr(aq)$	$\rightarrow$	$2HBr(g)$	$\Delta H^\circ = 160$ kJ
$Br_2(\ell)$	$\rightarrow$	$Br_2(aq)$	$\Delta H^\circ = -4.2$ kJ

Add all of the above to get;

$H_2(g) + Br_2(\ell)$	$\rightarrow$	$2HBr(g)$;	$\Delta H = -86$ kJ

Now divide this equation by two to give the thermochemical equation for the formation of 1 mol of HBr(g):

$1/2\ H_2(g) + 1/2\ Br_2(\ell)$	$\rightarrow$	$HBr(g)$;	$\Delta H = -43$ kJ

Comparing this value to the ΔH_f° value listed in Appendix E and at the outset of this problem, we see that this experimental data indicates a value that is close to the reported value.

6.100

$1/2HCHO_2(\ell) + 1/2H_2O(\ell) \rightarrow 1/2CH_3OH(\ell) + 1/2O_2(g)$	$\Delta H^\circ = +206$kJ
$1/2CO(g) + H_2(g) \rightarrow 1/2CH_3OH(\ell)$	$\Delta H^\circ = -64$ kJ
$1/2HCHO_2(\ell) \rightarrow 1/2CO(g) + 1/2H_2O(\ell)$	$\Delta H^\circ = -17$ kJ

Add these together:

$HCHO_2(\ell) + H_2(g) \rightarrow CH_3OH(\ell) + 1/2O_2(g)$	$\Delta H^\circ = +125$ kJ

6.102   Add together the fourth equation, the first equation, the second equation and the reverse of the third equation:

$4CuS(s) + 2CuO(s) \rightarrow 3Cu_2S(s) + SO_2(g)$	$\Delta H^\circ = -13.1$ kJ
$2Cu(s) + O_2(g) \rightarrow 2CuO(s)$	$\Delta H^\circ = -155$ kJ
$Cu(s) + S(s) \rightarrow CuS(s)$	$\Delta H^\circ = -53.1$ kJ
$SO_2(g) \rightarrow S(s) + O_2(g)$	$\Delta H^\circ = +297$ kJ

The net reaction is:

$3CuS(s) + 3Cu(s) \rightarrow 3Cu_2S(s)$	$\Delta H^\circ = +76$ kJ

We want 1/3 of this:

$CuS(s) + Cu(s) \rightarrow Cu_2S(s)$	$\Delta H^\circ = +25$ kJ

Practice Exercises

P7.1 $\quad v = \dfrac{c}{\lambda} = \dfrac{3.00 \times 10^8 \text{ m s}^{-1}}{550 \times 10^{-9} \text{ m}} = 5.45 \times 10^{14} \text{ s}^{-1}$

P7.2 $\quad \lambda = \dfrac{c}{v} = \dfrac{2.9979 \times 10^8 \text{ m s}^{-1}}{104.3 \times 10^6 \text{ s}^{-1}} = 2.874 \text{ m}$

P7.3

$$\frac{1}{\lambda} = 109{,}678 \text{ cm}^{-1} \times \left(\frac{1}{2^2} - \frac{1}{3^2} \right) = 109{,}678 \text{ cm}^{-1} \times (0.2500 - 0.1111)$$

$$\frac{1}{\lambda} = 1.523 \times 10^4 \text{ cm}^{-1}$$

$\lambda = 6.566 \times 10^{-5}$ cm = 656.6 nm, which is red.

P7.4   When n = 3, ℓ = 0, 1, 2. Thus we have s, p and d subshells.
            When n = 4, ℓ = 0, 1, 2, 3. Thus we have s, p, d and f subshells.

P7.5   For the g subshell, ℓ = 4 and there are 9 possible values of m_ℓ; −4, −3, −2, −1, 0, +1, +2, +3, +4. There are therefore 9 orbitals.

P7.6   (a)   Mg: $1s^2 2s^2 2p^6 3s^2$
            (b)   Ge: $1s^2 2s^2 2p^6 3s^2 3p^6 3d^{10} 4s^2 4p^2$
            (c)   Cd: $1s^2 2s^2 2p^6 3s^2 3p^6 3d^{10} 4s^2 4p^6 4d^{10} 5s^2$
            (d)   Gd: $1s^2 2s^2 2p^6 3s^2 3p^6 3d^{10} 4s^2 4p^6 4d^{10} 4f^7 5s^2 5p^6 5d^1 6s^2$

P7.7

(a) Na: (⥮) (⥮) (⥮)(⥮)(⥮) (↑) ( )( )( ) ( ) ( )( )( )( )( )
                     1s   **2s**   **2p**   **3s**   **3p**   **4s**   **3d**

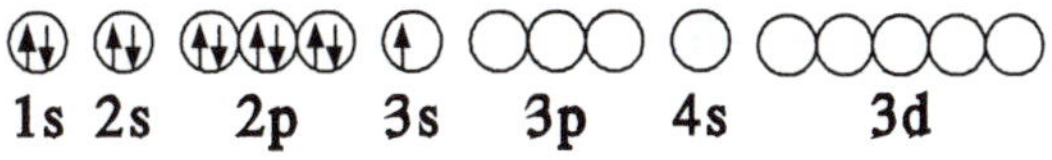

(b) S: (⥮) (⥮) (⥮)(⥮)(⥮) (⥮) (⥮)(↑)(↑) ( ) ( )( )( )( )( )
                     1s   **2s**   **2p**   **3s**   **3p**   **4s**   **3d**

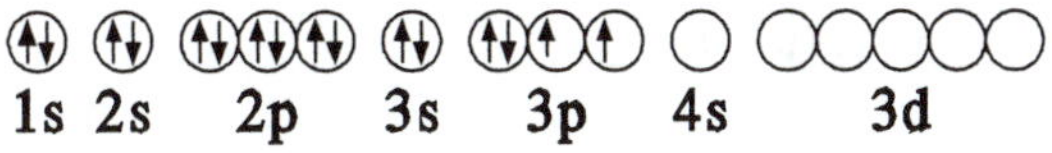

(c) Fe: (⥮) (⥮) (⥮)(⥮)(⥮) (⥮) (⥮)(⥮)(⥮) (⥮) (⥮)(↑)(↑)(↑)(↑)
                     1s   **2s**   **2p**   **3s**   **3p**   **4s**   **3d**

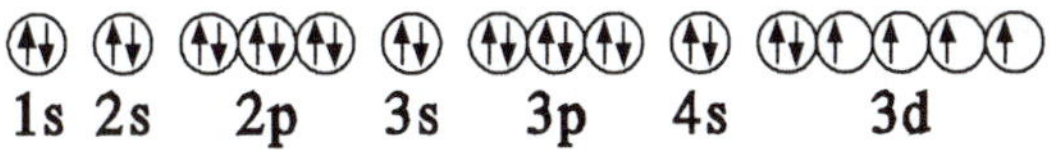

P7.8   (a)   P: $[\text{Ne}]3s^2 3p^3$

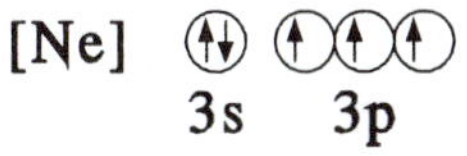

            [Ne]   (⥮) (↑)(↑)(↑)
                      3s   **3p**

            3 unpaired electrons

(b)   Sn: $[Kr]4d^{10}5s^25p^2$

[Kr] ⊕⊕⊕⊕⊕ ⊕ ⊕⊕○
            4d   **5s**   **5p**

2 unpaired electrons

P7.9   (a)   Se: $4s^24p^4$
       (b)   Sn: $5s^25p^2$
       (c)   I: $5s^25p^5$

P7.10 (a)   Sn
       (b)   Ga
       (c)   Cr
       (d) S^{2-}

P7.11 (a)   Be
       (b)   C

Review Problems

7.66 $v = \dfrac{c}{\lambda} = \dfrac{3.00 \times 10^8 \text{ m/s}}{430 \times 10^{-9} \text{ m}} = 6.98 \times 10^{14} \text{ s}^{-1} = 6.98 \times 10^{14} \text{ Hz}$

7.68   295 nm = 295×10^{-9} m

       $v = \dfrac{c}{\lambda} = \dfrac{3.00 \times 10^8 \text{ m/s}}{295 \times 10^{-9} \text{ m}} = 1.02 \times 10^{15} \text{ s}^{-1} = 1.02 \times 10^{15} \text{ Hz}$

7.70   101.1 MHz = 101.1×10^6 Hz = $101.1 \times 10^6 \text{ s}^{-1}$

       $\lambda = \dfrac{c}{v} = \dfrac{3.00 \times 10^8 \text{ m/s}}{101.1 \times 10^6 \text{ s}^{-1}} = 2.98 \text{ m}$

7.72 $\lambda = \dfrac{c}{v} = \dfrac{3.00 \times 10^8 \text{ m/s}}{60 \text{ s}^{-1}} = 5.0 \times 10^6 \text{ m} = 5.0 \times 10^3 \text{ km}$

7.74   E = hv = 6.63×10^{-34} J s x $4.0 \times 10^{14} \text{ s}^{-1} = 2.7 \times 10^{-19}$ J

       $\dfrac{\# \text{ J}}{\text{mol}} = \left(\dfrac{2.7 \times 10^{-19} \text{ J}}{1 \text{ photon}}\right)\left(\dfrac{6.022 \times 10^{23} \text{ photons}}{1 \text{ mol}}\right) = 1.6 \times 10^5 \text{ J mol}^{-1}$

7.76   (a)   violet

       (b) $v = c/\lambda = 3.00 \times 10^8 \text{ m s}^{-1}$ 410.3×10^{-9} m = $7.31 \times 10^{14} \text{ s}^{-1}$

       (c)   E = hv = 6.63×10^{-34} J s x $7.31 \times 10^{14} \text{ s}^{-1} = 4.85 \times 10^{-19}$ J

7.78

$$\frac{1}{\lambda} = 109,678 \, \text{cm}^{-1} \times \left(\frac{1}{3^2} - \frac{1}{6^2} \right) = 109,678 \, \text{cm}^{-1} \times (0.1111 - 0.02778)$$

$$\frac{1}{\lambda} = 9.140 \times 10^3 \, \text{cm}^{-1}$$

$\lambda = 1.094 \times 10^{-4}$ cm = 1090 nm, which is not in the visible region.

7.80

$$\frac{1}{\lambda} = 109,678 \, \text{cm}^{-1} \times \left(\frac{1}{4^2} - \frac{1}{10^2} \right)$$

$$\frac{1}{\lambda} = 5.758 \times 10^3 \, \text{cm}^{-1}$$

$\lambda = 1.737 \times 10^{-6}$ m this is infrared

7.82   (a)   p   (b)   f

7.84   (a) $n = 3$, $\ell = 0$ (b) $n = 5$, $\ell = 2$

7.86   0, 1, 2, 3, 4, 5

7.88   (a) $m_\ell = 1, 0,$ or -1 (b) $m_\ell = 3, 2, 1, 0, -1, -2,$ or -3

7.90   When $m_\ell = -4$ the minimum value of ℓ is 4 and the minimum value of n is 5.

7.92

n	ℓ	m_ℓ	m_s
2	1	−1	+1/2
2	1	−1	−1/2
2	1	0	+1/2
2	1	0	−1/2
2	1	+1	+1/2
2	1	+1	−1/2

7.94   21 electrons have $\ell = 1$, 20 electrons have $\ell = 2$

7.96   (a)   S $1s^2 2s^2 2p^6 3s^2 3p^4$
        (b)   K $1s^2 2s^2 2p^6 3s^2 3p^6 4s^1$
        (c)   Ti $1s^2 2s^2 2p^6 3s^2 3p^6 3d^2 4s^2$
        (d)   Sn $1s^2 2s^2 2p^6 3s^2 3p^6 4s^2 3d^{10} 4p^6 4d^{10} 5s^2 5p^2$

7.98   (a)   Mn is $[\text{Ar}]4s^2 3d^5$, $\therefore$ five unpaired electrons
        (b)   As is $[\text{Ar}] 3d^{10} 4s^2 4p^3$, $\therefore$ three unpaired electrons
        (c)   S is $[\text{Ne}]3s^2 3p^4$, $\therefore$ two unpaired electrons
        (d)   Sr is $[\text{Kr}]5s^2$, $\therefore$ zero unpaired electrons
        (e)   Ar is $1s^2 2s^2 2p^6 3s^2 3p^6$, $\therefore$ zero unpaired electrons

7.100   (a)   Mg is $1s^2 2s^2 2p^6 3s^2$, zero unpaired electrons
        (b)   P is $1s^2 2s^2 2p^6 3s^2 3p^3$, three unpaired electrons
        (c)   V is $1s^2 2s^2 2p^6 3s^2 3p^6 3d^3 4s^2$, three unpaired electrons

7.102   (a) $[\text{Ar}] 3d^8 4s^2$
        (b) $[\text{Xe}]6s^1$
        (c) $[\text{Ar}] 3d^{10} 4s^2 4p^2$

(d)   [Ar] $3d^{10}4s^24p^5$

(e)   [Xe] $4f^{14}5d^{10}6s^26p^3$

7.104

(a) <u>Mg</u>:

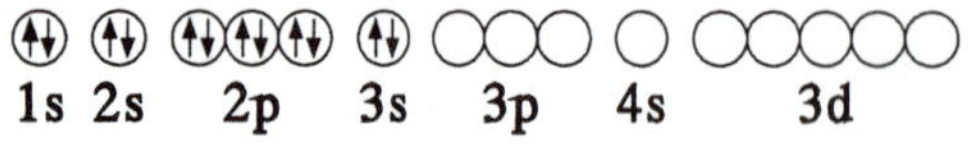

⊕ ⊕ ⊕⊕⊕ ⊕ ○○○ ○ ○○○○○
1s 2s   2p   3s 3p 4s   3d

(b) <u>Ti</u>:

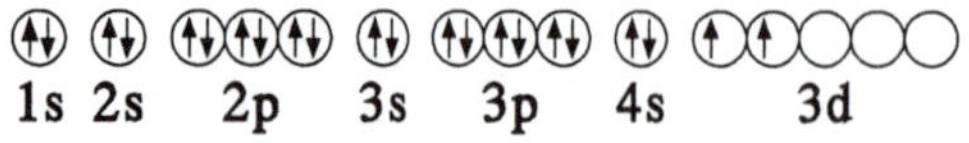

⊕ ⊕ ⊕⊕⊕ ⊕ ⊕⊕⊕ ⊕ ⊕⊕○○○
1s 2s   2p   3s   3p   4s   3d

7.106

(a) <u>Ni</u>: [Ar] ⊕ ⊕⊕⊕⊕⊕
                4s   3d

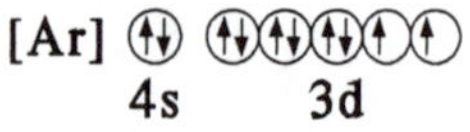

(b) <u>Cs</u>: [Xe] ⊕
                6s

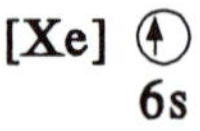

(c) <u>Ge</u>: [Ar] ⊕ ⊕⊕⊕⊕⊕   ⊕⊕○
                4s   3d   4p

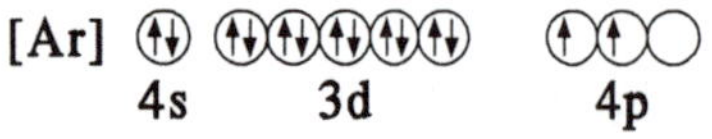

(d) <u>Br</u>: [Ar] ⊕ ⊕⊕⊕⊕⊕   ⊕⊕⊕
                4s   3d   4p

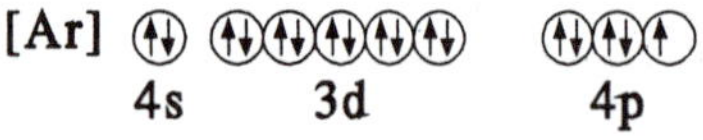

7.108 The value corresponds to the row in which the element resides:
      (a)   5   (b)   4   (c)   4   (d)   6

7.110 (a) Na $3s^1$ (b) Al $3s^23p^1$ (c) Ge $4s^24p^2$ (d) P $3s^23p^3$

7.112

(a) <u>Na</u>:   ⊕
                    3s

(b) <u>Al</u>:   ⊕ ⊕○○
                    3s   3p

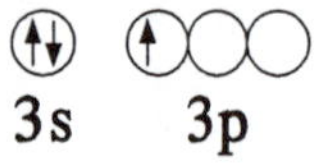

(c) Ge: (↑↓) (↑)(↑)( )

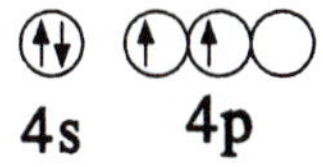

                 4s   4p

(d) P: (↑↓) (↑)(↑)(↑)

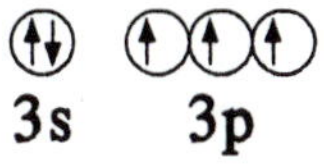

                 3s   3p

7.114   (a)   1   (b)   6   (c)   7

7.116   (a)   Na   (b)   Sb

7.118   Sb is a bit larger than Sn, according to Figure 7.25.

7.120   Cations are generally smaller than the corresponding atom, and anions are generally larger than the corresponding atom:
       (a)   Na   (b) Co^{2+} (c) Cl^-

7.122   (a)   C   (b)   O   (c)   Cl

7.124   (a)   Cl   (b)   Br

7.126   Mg

Additional Exercises

7.128   We proceed by calculating the wavelength of a single photon:

$$E = \frac{hc}{\lambda} = \frac{(6.626 \times 10^{-34}\ J\,s)(3.00 \times 10^8\ m/s)}{(3.00 \times 10^{-3}\ m)} = 6.63 \times 10^{-23}\ J$$

It requires 4.184 J to increase the temperature of 1.00 g of water by 1 Celsius degree. So, #photons = 4.184J ÷ 6.63 x 10^{-23} J/photon = 6.32 x 10^{22} photons

7.130   This corresponds to the special case in the Rydberg equation for which $n_1 = 1$ and $n_2 = \infty$.

$$\frac{1}{\lambda} = 109{,}678\ cm^{-1} \times \left(\frac{1}{1^2} - \frac{1}{\infty^2}\right) = 109{,}678\ cm^{-1} \times (1 - 0)$$

$$\frac{1}{\lambda} = 109{,}678\ cm^{-1}$$

$$\lambda = 9.12 \times 10^{-6}\ cm = 91.2\ nm.$$

7.132

$$E = 1/2\, m v^2$$

$$v = \sqrt{\frac{2E}{m}} \text{ first determine E for a single particle}$$

$$E = (2080 \times 10^3 \text{ J/mol})\left(\frac{1 \text{ mole}}{6.022 \times 10^{23} \text{ particles}}\right) = 3.454 \times 10^{-18} \text{ J/particle}$$

$$v = \sqrt{\frac{2(3.454 \times 10^{-18} \text{ J})}{9.109 \times 10^{-31} \text{ kg}}} = 2.754 \times 10^6 \text{ m/sec}$$

7.134 To solve this problem use $E = hc/\lambda$, where λ is our unknown quantity.

$$\lambda = \frac{hc}{E} = \frac{\left(6.626 \times 10^{-34} \text{ J s}\right)\left(3.00 \times 10^8 \text{ m/s}\right)}{\left(328 \times 10^3 \text{ J/mol}\right)}\left(6.02 \times 10^{23} \text{ photons/mol}\right)$$

$$= 3.65 \times 10^{-7} \text{ m} = 365 \text{ nm}$$

7.136 (a) The s shell is not completely filled.

(b) Using the shorthand notation of [Kr], we imply that the 4s, 3d and 4p shells are filled but the rest of the configuration negates this assumption.

(c) There is nothing wrong.

(d) The s shell should be completely filled. If moving an electron from the s subshell would have filled the d subshell, then we could have this case but the d subshell is still two electrons short of filling.

7.138 14

7.140 (a) This corresponds to the special case in the Rydberg equation for which $n_1 = 1$ and $n_2 = \infty$. For a single atom, we have:

$$\frac{1}{\lambda} = 109{,}678 \text{ cm}^{-1} \times \left(\frac{1}{1^2} - \frac{1}{\infty^2}\right) = 109{,}678 \text{ cm}^{-1} \times (1 - 0)$$

$$\frac{1}{\lambda} = 109{,}678 \text{ cm}^{-1}$$

$$\lambda = 9.12 \times 10^{-6} \text{ cm} = 91.2 \text{ nm}.$$

Converting to energy, we have:

$$E = hc/\lambda$$

$$E = \frac{hc}{\lambda} = \frac{\left(6.626 \times 10^{-34} \text{ J s}\right)\left(3.00 \times 10^8 \text{ m/s}\right)}{\left(91.2 \times 10^{-9} \text{ m}\right)} = 2.18 \times 10^{-18} \text{ J}$$

(b) Conversion to kJ/mol gives us:

$$\# \text{ kJ/mole} = \left(\frac{2.18 \times 10^{-18} \text{ J}}{\text{photon}}\right)\left(\frac{1 \text{ kJ}}{1000 \text{ J}}\right)\left(\frac{6.02 \times 10^{23} \text{ photons}}{\text{mole}}\right)$$

$$= 1.31 \times 10^3 \text{ kJ/mol}$$

7.142   We simply reverse the electron affinities of the corresponding ions.

(a) $F^-(g) \rightarrow F(g) + e^-$, $H° = 328$ kJ/mol

(b) $O^-(g) \rightarrow O(g) + e^-$, $H° = 141$ kJ/mol

(c) $O^{2-}(g) \rightarrow O^-(g) + e^-$, $H° = -844$ kJ/mol

The last of these is exothermic, meaning that loss of an electron from the oxide ion is favorable from the standpoint of enthalpy.

<u>Practice Exercises</u>

P8.1   Cr: [Ar] $3d^4 4s^2$
       (a) Cr^{2+}: [Ar]$3d^4$
       (b) Cr^{3+}: [Ar]$3d^3$
       (c) Cr^{6+}: [Ar]

P8.2

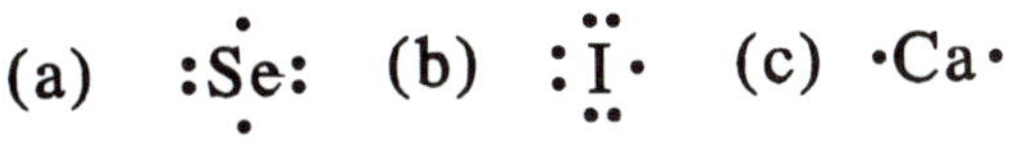

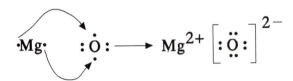

(a) :S̈e: **(b)** :Ï· **(c)** ·Ca·

P8.3

·Mg·   :Ö: ⟶ Mg^{2+} $\left[:\ddot{O}: \right]^{2-}$

P8.4   (a)   Br is more electronegative and carries the partial negative charge.
       (b)   Cl is more electronegative and carries the partial negative charge.
       (c)   Cl is more electronegative and carries the partial negative charge.

P8.5

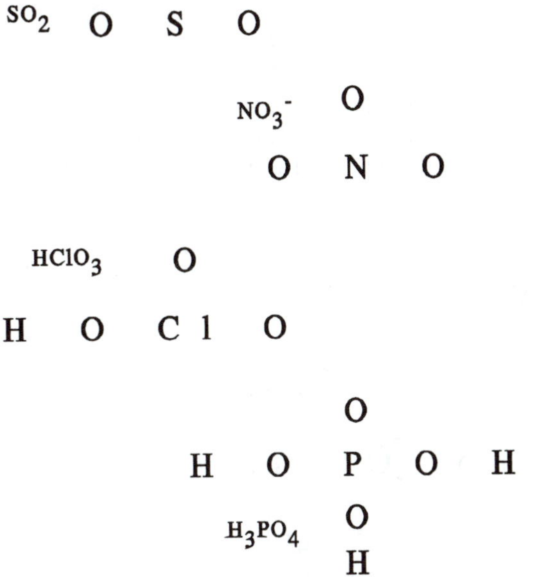

SO_2 O   S   O

NO_3^- O

                 O   N   O

$HClO_3$ O

       H   O   C   l   O

                            O

               H   O   P   O   H

H_3PO_4 O

                       H

P8.6 SO_2 has 18 valence electrons
       PO_4^{3-} has 32 valence electrons
       NO^+ has 10 valence electrons

P8.7

:F̈—Ö—F̈:   Ö = S—Ö:

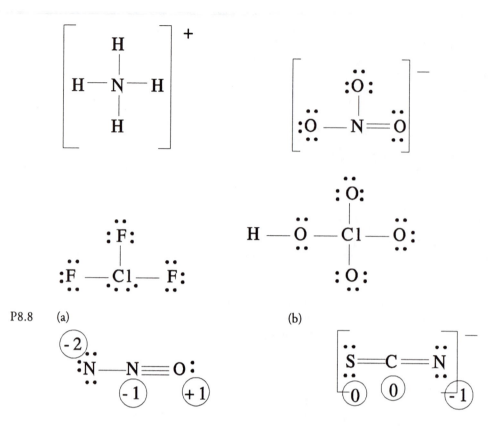

P8.8   (a)   (b)

P8.9   In each of these problems, we try to minimize the formal charges in order to determine the preferred Lewis structure. This frequently means that we will violate the octet rule by expanding the octet. Of course, we can only do this for atoms beyond the second period as the atoms in the first and second periods will never expand the octet.

(a)

(b)

(c)

P8.10 (a)

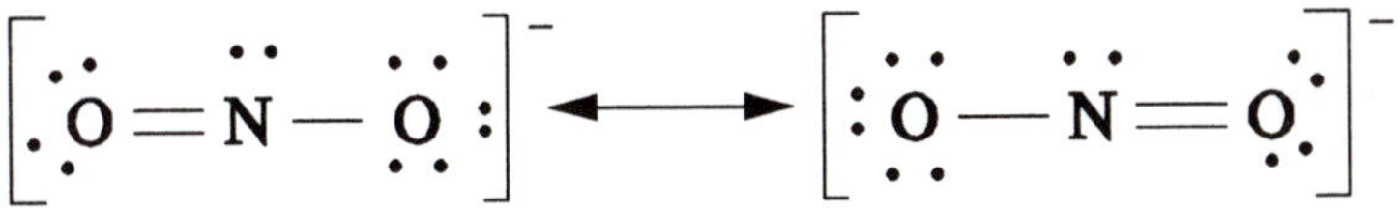

$$\left[\,\ddot{O}=\ddot{N}-\ddot{O}:\,\right]^{-} \longleftrightarrow \left[\,:\ddot{O}-\ddot{N}=\ddot{O}\,\right]^{-}$$

(b)

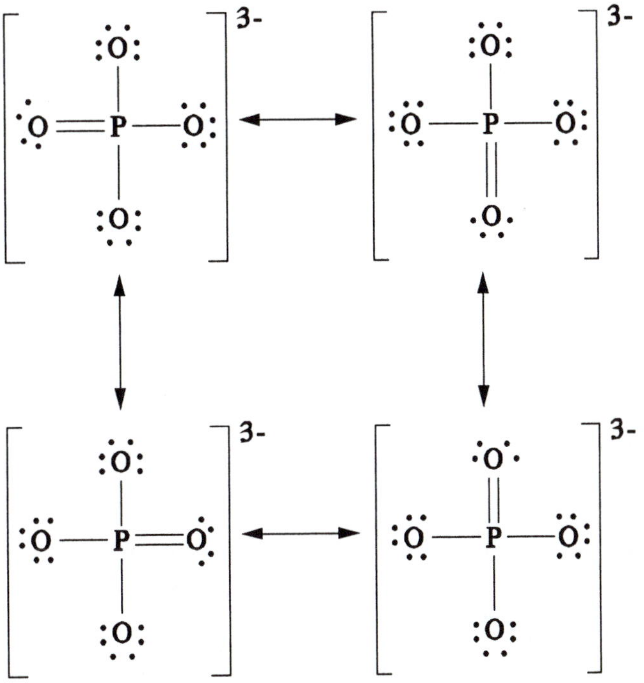

$$\left[\begin{array}{c}:\ddot{O}:\\ :\ddot{O}=P-\ddot{O}:\\ :\ddot{O}:\end{array}\right]^{3-} \longleftrightarrow \left[\begin{array}{c}:\ddot{O}:\\ :\ddot{O}-P-\ddot{O}:\\ .\ddot{O}.\end{array}\right]^{3-}$$

$$\updownarrow \qquad\qquad \updownarrow$$

$$\left[\begin{array}{c}:\ddot{O}:\\ :\ddot{O}-P=\ddot{O}\\ :\ddot{O}:\end{array}\right]^{3-} \longleftrightarrow \left[\begin{array}{c}.\ddot{O}.\\ :\ddot{O}-P-\ddot{O}:\\ :\ddot{O}:\end{array}\right]^{3-}$$

P8.11

$$H^{+} + \left[\,:\ddot{O}-H\,\right]^{-}$$

$$H-\ddot{O}-H$$

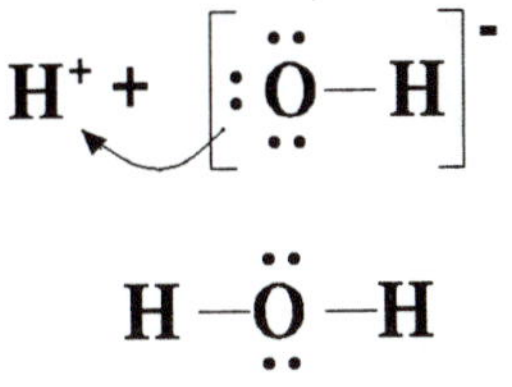

Review Problems

8.57 Magnesium loses two electrons:
$$Mg \rightarrow Mg^{2+} + 2e^{-}$$
$$[Ne]3s^2 \quad [Ne]$$

Bromine gains an electron:
$$Br + e^{-} \rightarrow Br^{-}$$
$$[Ar]3d^{10}4s^2 4p^5 \quad \rightarrow \quad [Kr]$$

To keep the overall change of the molecule neutral, two Br^- ions combine with one Mg^{2+} ion to form $MgBr_2$:
$$Mg^{2+} + 2Br^{-} \rightarrow MgBr_2$$

8.59 Pb^{2+}: $[Xe]\,4f^{14}5d^{10}6s^2$

Pb^{4+}: [Xe]4f^{14}5d^{10}

8.61   Mn^{3+}:   [Ar]3d^4 4 unpaired electrons

8.63

(a)   ·S̈i·   (b)   ·S̈b̈·

(c)   ·Ba·   (d)   ·Äl·

(e)   :S̈·

8.65

(a)   [K]$^+$ (b)   [Al]$^{3+}$

(c)   [:S̈:]$^{2-}$ (d)   [:S̈i:]$^{4-}$

(e)   [Mg]$^{2+}$

8.67

(a)

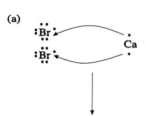

2[:B̈r:]$^-$ +   [Ca]$^{2+}$

(b)

2 [Al]$^{3+}$ +   3 [:Ö:]$^{2-}$

(c)

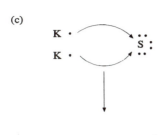

$2[K]^+ + [:\overset{\cdot\cdot}{\underset{\cdot\cdot}{S}}:]^{2-}$

8.69   Let q be the amount of energy released in the formation of 1 mol of H_2 molecules from H atoms: 435 kJ/mol, the single bond energy for hydrogen.

q = specific heat $\times$ mass $\times \Delta$T
$\therefore$ mass = q ÷ (specific heat $\times \Delta$T)

$$\# \text{ g } H_2O = \frac{(435 \times 10^3 \text{ J})}{\left(4.184 \text{ } \frac{J}{g \text{ }^\circ C}\right)\left(100 \text{ }^\circ C - 25 \text{ }^\circ C\right)} = 1.4 \times 10^3 \text{ g}$$

8.71

$$E = h\nu = \frac{hc}{\lambda} \qquad \lambda = \frac{hc}{E}$$

$$E = (348 \times 10^3 \text{ J/mol})\left(\frac{1 \text{ mol}}{6.022 \times 10^{23} \text{ molecules}}\right) = 5.78 \times 10^{-19} \text{ J/molecule}$$

$$\lambda = (6.626 \times 10^{-34} \text{ Jsec})(3.00 \times 10^8 \text{ m/s})/5.78 \times 10^{-19} \text{ J} = 3.44 \times 10^{-7} \text{ m} = 344 \text{ nm}$$

ultraviolet

8.73

(a) $:\overset{\cdot\cdot}{\underset{\cdot\cdot}{Br}}\cdot \ + \ \cdot\overset{\cdot\cdot}{\underset{\cdot\cdot}{Br}}: \ \longrightarrow \ :\overset{\cdot\cdot}{\underset{\cdot\cdot}{Br}} - \overset{\cdot\cdot}{\underset{\cdot\cdot}{Br}}:$

(b) $2 \text{ H}\cdot \ + \ \cdot\overset{\cdot\cdot}{\underset{\cdot\cdot}{O}}\cdot \ \longrightarrow \ \text{H} - \overset{\cdot\cdot}{\underset{\cdot\cdot}{O}}: \atop \hspace{1.5em} |\atop \hspace{1.5em} \text{H}$

(c) $3 \text{ H}\cdot \ + \ \cdot\overset{\cdot\cdot}{\underset{\cdot}{N}}\cdot \ \longrightarrow \ \text{H} - \overset{\cdot\cdot}{N} - \text{H} \atop \hspace{2em} |\atop \hspace{2em} \text{H}$

8.75   (a)   We predict the formula H_2Se because selenium, being in Group VIA, needs only two additional electrons (one each from two hydrogen atoms) in order to complete its octet.

(b)   Arsenic, being in Group VA, needs three electrons from hydrogen atoms in order to complete its octet, and we predict the formula H_3As.

(c)   Silicon is in Group IVB, and it needs four electrons (and hence four hydrogen atoms) to complete its octet: SiH_4.

8.77   Here we choose the atom with the smaller electronegativity:
(a)   S   (b)   Si   (c)   Br   (d)   C

8.79   Here we choose the linkage that has the greatest difference in electronegativities between the atoms of the bond: N—S.

8.81

(a)

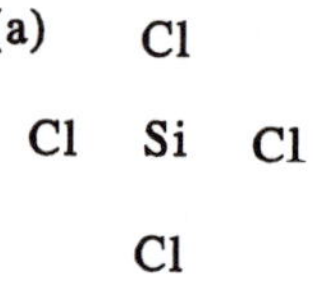

Cl
Cl Si Cl
Cl

(b)

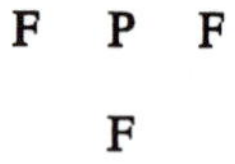

F P F
F

(c)

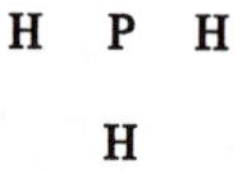

H P H
H

(d)

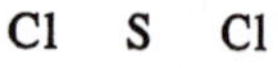

Cl S Cl

8.83 (a) 32 (b) 26 (c) 8 (d) 20

8.85

(a)

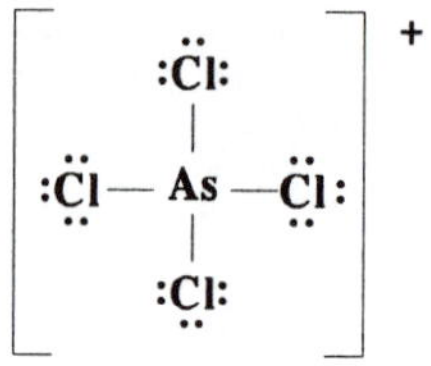

$$\left[\begin{array}{c} :\ddot{C}l: \\ | \\ :\ddot{C}l-As-\ddot{C}l: \\ | \\ :\ddot{C}l: \end{array}\right]^{+}$$

(b)

$$\left[:\ddot{O}-\ddot{C}l-\ddot{O}:\right]^{-}$$

(c)

$$H-\ddot{O}-\ddot{N}=\ddot{O}$$

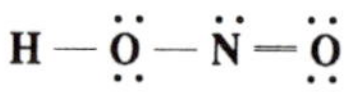

(d)

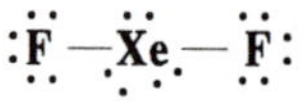

$$:\ddot{F}-\ddot{X}e-\ddot{F}:$$

8.87

(a)

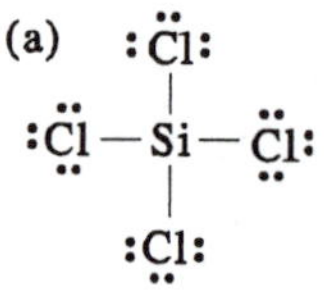

$$\begin{array}{c} :\ddot{C}l: \\ | \\ :\ddot{C}l-Si-\ddot{C}l: \\ | \\ :\ddot{C}l: \end{array}$$

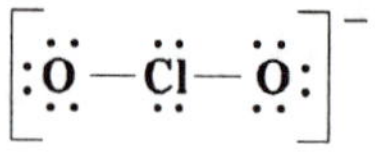

(b)

$$\begin{array}{c} :\ddot{F}-\ddot{P}-\ddot{F}: \\ | \\ :\ddot{F}: \end{array}$$

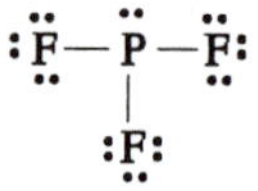

(c)

$$H-\ddot{P}-H$$
$$|$$
$$H$$

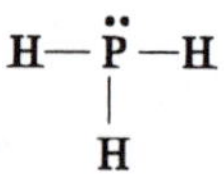

(d)

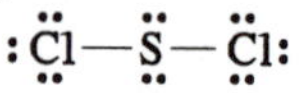

$$:\ddot{C}l-\ddot{S}-\ddot{C}l:$$

8.89

(a)

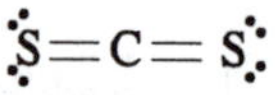

$$\ddot{S}=C=\ddot{S}$$

(b)

$$\left[:C\equiv N:\right]^{-}$$

8.91

(a)

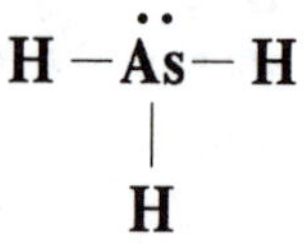

$$H-\ddot{A}s-H$$
$$|$$
$$H$$

(b)

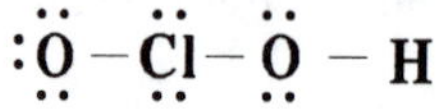

$$:\ddot{O}-\ddot{C}l-\ddot{O}-H$$

(c)

$$H-\ddot{O}-\ddot{S}e-\ddot{O}-H$$
$$:\ddot{O}: :\ddot{O}:$$

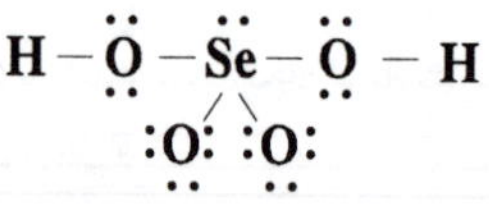

(d)

8.93

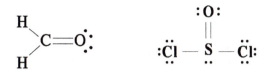

8.95

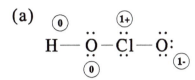

(a)

(b)

(c)

8.97

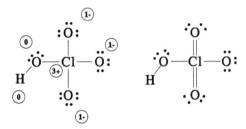

8.99 The formal charges on all of the atoms of the left structure are zero, therefore, the potential energy of this molecule is lower and it is more stable.

8.101 The average bond order is 4/3

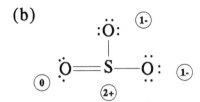

8.103 The Lewis structure for NO_3^- is given in the answer to practice exercise 7, and that for NO_2^- is given in the answer to review exercise 8.92.

Resonance causes the average number of bonds in each N—O linkage of NO_3^- to be 1.33. Resonance causes the average number of electron pair bonds in each linkage of NO_2^- to be 1.5. We conclude that the N—O bond in NO_2^- should be shorter than that in NO_3^-.

8.105

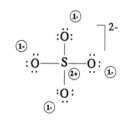

These are not preferred structures, because in each Lewis diagram, one oxygen bears a formal charge of +1 whereas the other bears a formal charge of –1. The structure with the formal charges of zero has a lower potential energy and is more stable.

8.107   The Lewis structure that is obtained using the rules of Figure 8.9 is:

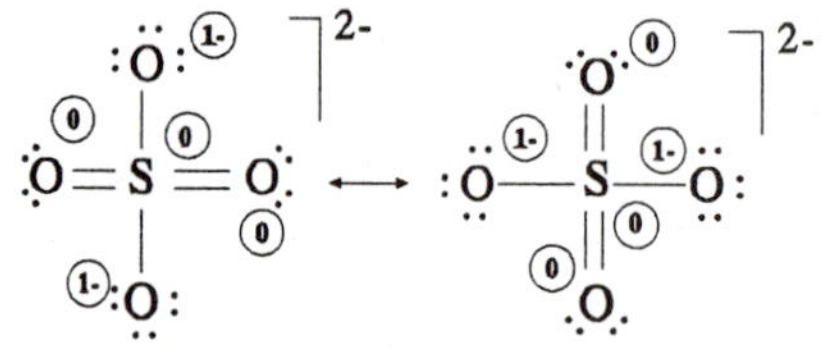

Formal charges are indicated. The average bond order is 1.5.

8.109

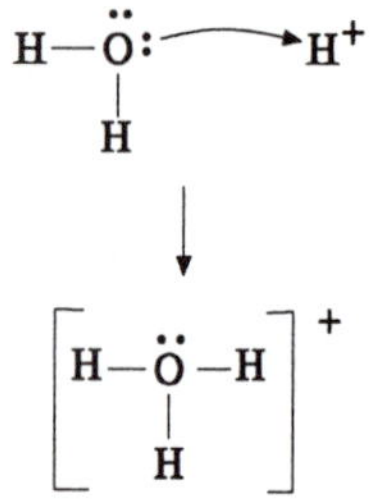

Additional Exercises

8.111 $Na(g) \rightarrow Na^+(g) + e^-$ IE = 496 kJ/mol
        $Cl(g) + e^- \rightarrow Cl^-(g)$ EA = –348 kJ/mol
        $Na(g) + Cl(g) \rightarrow Na^+(g) + Cl^-(g)$ ΔH = 148 kJ/mol

        $Na(g) \rightarrow Na^+(g) + e^-$ IE = 496 kJ/mol
        $Na^+(g) \rightarrow Na^{2+}(g) + e^-$ IE = 4563 kJ/mol
        $2Cl(g) + 2e^- \rightarrow 2Cl^-$ EA = 2(–348kJ/mol)
        $Na(g) + 2Cl(g) \rightarrow Na^{2+}(g) + 2Cl^-(g)$ ΔH = 4.36 x 10^3 kJ/mol

In order for $NaCl_2$ to be more stable than NaCl, the lattice energy should be almost 30 times larger 436 kJ/148 kJ = 29.5.

8.113

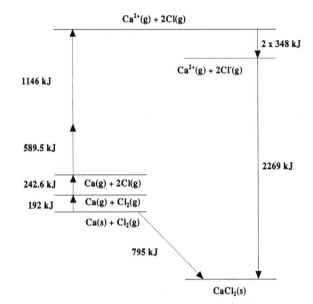

Ca²⁺(g) + 2Cl(g)

2 x 348 kJ

Ca²⁺(g) + 2Cl⁻(g)

1146 kJ

589.5 kJ

2269 kJ

242.6 kJ — Ca(g) + 2Cl(g)

192 kJ — Ca(g) + Cl₂(g)

Ca(s) + Cl₂(g)

795 kJ

CaCl₂(s)

8.118   (a)   Carbon nearly always makes four bonds and this structure has only 3. Additionally, the formal charge on the carbon atom is −1.

(b)   The formal charge on the oxygen atom bonded to the hydrogen is +1 which is unlikely since it is the most electronegative element in the molecule.

(c)   Carbon never makes more than four bonds.

8.120

8.123

$$:\ddot{C}l - \ddot{S} - \ddot{S} - \ddot{C}l:$$

Practice Exercises

P9.1   SbCl₅ should have a trigonal bipyramidal shape (Figure 9.2) because, like PCl₅, it has five electron pairs around the central atom.

P9.2   In SO_3^{2-}, there are three bond pairs and one lone pair of electrons at the sulfur atom, and as shown in Figure 9.3, this ion has a pyramidal shape.

In XeO_4, there are four bond pairs of electrons around the Xe atom, and as shown in Figure 9.3, this molecule is tetrahedral.

In OF_2, there are two bond pairs and two lone pairs of electrons around the oxygen atom, and as shown in Figure 9.3, this molecule is bent.

P9.3   Carbonate ion, CO_3^{2-}, is a planar, triangular ion, having no lone pairs on the central carbon. It has the same geometry as SO_3.

P9.4   (a) SF_6 is octahedral, and it is not polar.
        (b) SO_2 is bent, and it is polar.
        (c)   BrCl is polar because there is a difference in electronegativity between Br and Cl.
        (d) AsH_3, like NH_3, is pyramidal, and it is polar.
        (e) CF_2Cl_2 is polar, because there is a difference in electronegativity between F and Cl.

P9.5   The H–Cl bond is formed by the overlap of the half–filled 1s atomic orbital of a H atom with the half–filled 3p valence orbital of a Cl atom:

Cl atom in HCl (x = H electron):

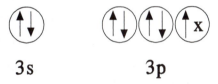

3s   3p

The overlap that gives rise to the H–Cl bond is that of a 1s orbital of H with a 3p orbital of Cl:

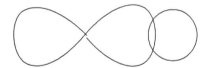

P9.6   The half–filled 1s atomic orbital of each H atom overlaps with a half–filled 3p atomic orbital of the P atom, to give three P–H bonds. This should give a bond angle of 90°.

P atom in PH₃ (x = H electron):

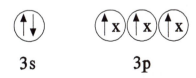

3s   3p

The orbital overlap that forms the P–H bond combines a 1s orbital of hydrogen with a 3p orbital of phosphorus (note: only half of each p orbital is shown):

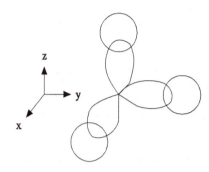

P9.7   Since there are five bonding pairs of electrons on the central arsenic atom, we choose sp^3d hybridization for the As atom. Each of arsenic's five sp^3d hybrid orbitals overlaps with a 3p atomic orbital of a chlorine atom to form a total of five As–Cl single bonds. Four of the 3d atomic orbitals of As remain unhybridized.

P9.8   (a) sp^3 (b) sp^3d

P.9   (a) sp^3 (b) sp^3d

P9.10 sp^3d^2, since six atoms are bonded to the central atom.

P atom in PCl_6^- (x = Cl electron):

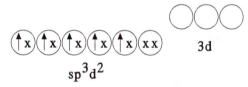

3d

sp^3d^2

The ion is octahedral because six atoms and no lone pairs surround the central atom.

P9.11   NO has 11 valence electrons, and the MO diagram is similar to that shown in Table 9.1 for O_2, except that one fewer electron is employed at the highest energy level

$\sigma^*_{2p_z}$ ··········   ⟮↑↓⟯

$\pi^*_{2p_x}, \pi^*_{2p_y}$ ··········   ⟮↑⟯   ◯

π_{2p_x}, π_{2p_y} ··········   ⟮↑↓⟯   ⟮↑↓⟯

σ_{2p_x} ··········   ⟮↑↓⟯

σ^*_{2s} ··········   ⟮↑↓⟯

σ_{2s} ··········   ⟮↑↓⟯

The bond order is calculated to be 5/2:

$$\text{Bond Order} = \frac{\left(8 \text{ bonding e}^-\right) - \left(3 \text{ antibonding e}^-\right)}{2} = \frac{5}{2}$$

Review Problems

9.56   (a)   nonlinear   (b)   trigonal bipyramidal
       (c)   trigonal pyramidal   (d)   trigonal pyramidal
       (e)   nonlinear

9.58   (a)   tetrahedral   (b)   square planar
       (c)   octahedral   (d)   tetrahedral
       (e)   linear

9.60   180°

9.62   The ones that are polar are (a), (b), and (c). The last two have symmetrical structures, and although individual bonds in these substances are polar bonds, the geometry of the bonds serves to cause the individual dipole moments of the various bonds to cancel one another.

9.64   The 1s atomic orbitals of the hydrogen atoms overlap with the mutually perpendicular p atomic orbitals of the selenium atom.

Se atom in H$_2$Se (x = H electron):

9.66

Atomic Be:

Hybridized Be:

(x = a Cl electron)

9.68   (a)   There are three bonds to the central Cl atom, plus one lone pair of electrons. The geometry of the electron pairs is tetrahedral so the Cl atom is to be sp^3 hybridized:

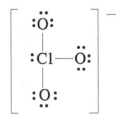

       (b)   There are three atoms bonded to the central sulfur atom, and no lone pairs on the central sulfur. The geometry of the electron pairs is that of a planar triangle, and the hybridization of the S atom is sp^2:

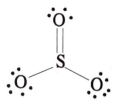

Two other resonance structures should also be drawn for SO_3.

(c)   There are two bonds to the central O atom, as well as two lone pairs. The O atom is to be sp^3 hybridized, and the geometry of the electron pairs is tetrahedral.

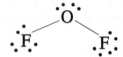

9.70   (a)   There are three bonds to As and one lone pair at As, requiring As to be sp^3 hybridized.

The Lewis diagram:

$$:\overset{..}{\underset{..}{Cl}}—\overset{..}{As}—\overset{..}{\underset{..}{Cl}}:$$

$$\underset{..}{\overset{|}{:\underset{..}{Cl}}:}$$

The hybrid orbital diagram for As:

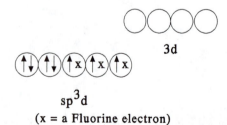

$$sp^3$$

(x = a Cl electron)

(b)   There are three atoms bonded to the central Cl atom, and it also has two lone pairs of electrons. The hybridization of Cl is thus sp^3d.

The Lewis diagram:

$$:\overset{..}{\underset{..}{F}}—\overset{..}{\underset{.}{Cl}}—\overset{..}{\underset{..}{F}}:$$

$$\overset{|}{:\underset{..}{F}:}$$

The hybrid orbital diagram for Cl:

○○○○
                              3d

sp^3d
(x = a Fluorine electron)

9.72   We can consider that this ion is formed by reaction of SbF₅ with F⁻. The antimony atom accepts a pair of electrons from fluoride:

Sb in SbF₆⁻:

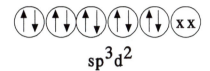

$$sp^3d^2$$

(xx = an electron pair from the donor F⁻)

9.74   (a)

N in the C=N system:

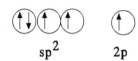

$$sp^2 \qquad 2p$$

(b)

sigma bond   pi bond

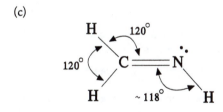

(c)

H   120°
120°   C══N
H   ~118°   H

9.76   Each carbon atom is sp² hybridized, and each C–Cl bond is formed by the overlap of an sp² hybrid of carbon with a p atomic orbital of a chlorine atom. The C=C double bond consists first of a C–C σ bond formed by "head on" overlap of sp² hybrids from each C atom. Secondly, the C=C double bond consists of a side–to–side overlap of unhybridized p orbitals of each C atom, to give one π bond. The molecule is planar, and the expected bond angles are all 120°.

9.78   1. sp³   2. sp   3. sp²   4. sp²

9.80   Here we pick the one with the higher bond order.

(a) O_2^+ (b) O_2 (c) N_2

9.82

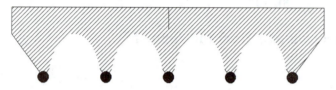

Empty 4p band

half-filled 4s band

The energy bands in solid potassium.

Additional Exercises

9.84   planar triangular

9.86   The normal C–C–C angle for an sp^3 hybridized carbon atom is 109.5°. The 60° bond angle in cyclopropane is much less than this optimum bond angle. This means that the bonding within the ring cannot be accomplished through the desirable "head on" overlap of hybrid orbitals from each C atom. As a result, the overlap of the hybrid orbitals in cyclopropane is less effective than that in the more normal, noncyclic propane molecule, and this makes the C–C bonds in cyclopropane comparatively weaker than those in the noncyclic molecule. We can also say that there is a severe "ring strain" in the molecule.

9.88   (a) PF_3 is a pyramidal molecule and uses sp^3 hybrid orbitals. The expected bond angle is 109.5°.
         (b)   Using the unhybidized p orbitals, we would anticipate a bond angle of 90°.
         (c)   The observed bond angle is almost exactly the average of the bond angles listed in parts (a) and (b) above. So, neither hybrid orbitals nor unhybridized atomic orbitals explain the observed bond angle.

9.90   (a)   The C–C single bonds are formed from head–to–head overlap of C atom sp^2 hybrids. This leaves one unhybridized atomic p orbital on each carbon atom, and each such atomic orbital is oriented perpendicular to the plane of the molecule.

         (b)   Sideways or π type overlap is expected between the first and the second carbon atoms, as well as between the third and the fourth carbon atoms. However, since all of these atomic p orbitals are properly aligned, there can be continuous π type overlap between all four carbon atoms.

         (c)   The situation described in part (b) is delocalized. We expect completely delocalized π type bonding among the carbon atoms.

         (d)   The bond is shorter because of the extra stability associated with the delocalization energy.

9.92   The arrangement of the atoms is trigonal bipyramidal.

Recall that the bond angle between equatorial atoms is 120°. The bond angle from the equatorial position to the axial position is 90°. Due to the smaller bond angles, the atoms in the axial positions create more repulsions. The structure with the least amount of total repulsion is preferred; the statement implies that the more

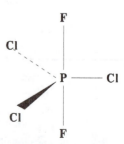

electronegative atoms create less repulsion, therefore the more electronegative atom should be placed in the axial position. Since fluorine is more electronegatove then chlorine, the F atoms will be in the axial positions and the Cl atoms will be in the equatorial positions. The molecule is non–polar.

9.94 The double bonds are predicted to be between S and O atoms. Hence, the Cl–S–Cl angle diminishes under the influence of the S=O double bonds.

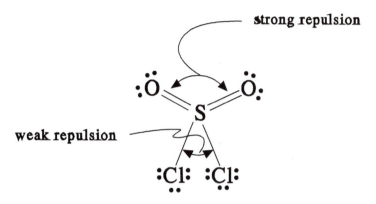

9.96 This is a π bond, since overlap takes place both above and below the bond axis.

Practice Exercises

P10.1   We will use Equation 10.3 to determine the length needed to measure 1 atm of pressure. (The density of Hg is found in Example 10.1.)

$$h_{H_2O} = \left(h_{Hg}\right)\left(\frac{d_{Hg}}{d_{H_2O}}\right) = (760 \text{ mm})\left(\frac{13.6 \text{ g mL}^{-1}}{1.00 \text{ g mL}^{-1}}\right) = 1.03 \times 10^4 \text{ mm}$$

$$\# \text{ ft} = \left(1.03 \times 10^4 \text{ mm}\right)\left(\frac{1 \text{ m}}{1000 \text{ mm}}\right)\left(\frac{100 \text{ cm}}{1 \text{ m}}\right)\left(\frac{1 \text{ in.}}{2.54 \text{ cm}}\right)\left(\frac{1 \text{ ft}}{12 \text{ in.}}\right) = 33.9 \text{ ft}$$

P10.2   In general the combined gas law equation is: $\dfrac{P_1V_1}{T_1} = \dfrac{P_2V_2}{T_2}$, and in particular, for this problem, we have:

$$P_2 = \frac{P_1V_1T_2}{T_1V_2} = \frac{(745 \text{ torr})(950 \text{ m}^3)(333.2 \text{ K})}{(1150 \text{ m}^3)(298.2 \text{ K})} = 688 \text{ torr}$$

P10.3   Since volume is to decrease, pressure must increase, and we multiply the starting pressure by a volume ratio that is larger than one. Also, since $P_1V_1 = P_2V_2$, we can solve for P_2:

$$P_2 = \frac{P_1V_1}{V_2} = \frac{(740 \text{ torr})(880 \text{ mL})}{(870 \text{ mL})} = 750 \text{ torr}$$

P10.4   We will use the ideal gas law. First we need to know the number of moles of nitrogen:

$$\# \text{ mol N}_2 = (0.245 \text{ g N}_2)\left(\frac{1 \text{ mol N}_2}{28.02 \text{ g N}_2}\right) = 8.74 \times 10^{-3} \text{ mol N}_2$$

$$V = \frac{nRT}{P} = \frac{\left(8.74 \times 10^{-3} \text{ mol}\right)\left(0.0821 \frac{\text{L atm}}{\text{mol K}}\right)(294 \text{ K})}{(750 \text{ torr})\left(\dfrac{1 \text{ atm}}{760 \text{ torr}}\right)} = 0.214 \text{ L} = 214 \text{ mL}$$

P10.5   Since PV = nRT, then n = PV/RT

$$n = \frac{PV}{RT} = \frac{(685 \text{ torr})\left(\dfrac{1 \text{ atm}}{760 \text{ torr}}\right)(0.300 \text{ L})}{\left(0.0821 \frac{\text{L atm}}{\text{mol K}}\right)(300.2 \text{ K})} = 0.0110 \text{ moles gas}$$

$$\text{molar mass} = \frac{1.45 \text{ g}}{0.0110 \text{ mol}} = 132 \text{ g mol}^{-1}$$

The gas must be Xenon.

P10.6   In general PV = nRT, where n = mass formula mass. Thus

$$PV = \frac{\text{mass}}{\text{formula mass}} RT$$

We can rearrange this equation to get;

$$\text{formula mass} = \frac{DRT}{P}$$

$$\text{formula mass} = \frac{\left(5.60\,\text{g L}^{-1}\right)\left(0.0821\,\frac{\text{L atm}}{\text{mol K}}\right)\left(295.2\,\text{K}\right)}{\left(750\,\text{torr}\right)\left(\dfrac{1\,\text{atm}}{760\,\text{torr}}\right)} = 138\,\text{g mol}^{-1}$$

The empirical mass is 69 g mol^{-1}. The ratio of the molecular mass to the empirical mass is 138 g mol^{-1}/69 g mol^{-1} = 2. Therefore, the molecular formula is 2 times the empirical formula, i.e., P_2F_4.

P10.7   When gases are held at the same temperature and pressure, and dispensed in this fashion during chemical reactions, then they react in a ratio of volumes that is equal to the ratio of the coefficients (moles) in the balanced chemical equation for the given reaction. We can, therefore, directly use the stoichiometry of the balanced chemical equation to determine the combining ratio of the gas volumes:

$$\# \text{L O}_2 = \left(4.50\,\text{L CH}_4\right)\left(\frac{2\,\text{volume O}_2}{1\,\text{volume CH}_4}\right) = 9.00\,\text{L O}_2$$

P10.8   First lets determine the number of moles of CO_2 that are produced:

$$n = \frac{PV}{RT} = \frac{\left(738\,\text{torr}\right)\left(\dfrac{1\,\text{atm}}{760\,\text{torr}}\right)\left(0.250\,\text{L}\right)}{\left(0.0821\,\dfrac{\text{L atm}}{\text{mol·K}}\right)\left(296\,\text{K}\right)} = 9.99 \times 10^{-3}\,\text{moles CO}_2$$

The stoichiometry of the reaction indicates that one mole of Na_2CO_3(s) will produce one mole CO_2(g), so we will need to use 9.99 X 10^{-3} mol of Na_2CO_3(s).

$$\# \text{g Na}_2\text{CO}_3 = \left(9.99 \times 10^{-3}\,\text{mol Na}_2\text{CO}_3\right)\left(\frac{106\,\text{g Na}_2\text{CO}_3}{1\,\text{mol Na}_2\text{CO}_3}\right) = 1.06\,\text{g Na}_2\text{CO}_3$$

P10.9   We can determine the pressure due to the oxygen since $P_{total} = P_{N_2} + P_{O_2}$.

$P_{O_2} = P_{total} - P_{N_2} = 30.0\,\text{atm} - 15.0\,\text{atm} = 15.0\,\text{atm}$. We can now use the ideal gas law to determine the number of moles of O_2:

$$n = \frac{PV}{RT} = \frac{\left(15.0\,\text{atm}\right)\left(5.00\,\text{L}\right)}{\left(0.0821\,\dfrac{\text{L atm}}{\text{mol K}}\right)\left(298\,\text{K}\right)} = 3.06\,\text{moles O}_2$$

$$\# \text{g O}_2 = \left(3.06\,\text{moles O}_2\right)\left(\frac{32.0\,\text{g O}_2}{1\,\text{mol O}_2}\right) = 98.1\,\text{g O}_2$$

P10.10 First we find the partial pressure of nitrogen:

$P_{N_2} = P_{total} - P_{water} = 745\,\text{torr} - 12.79\,\text{torr} = 732\,\text{torr}$.

To calculate the volume of the nitrogen we can use the combined gas law

$$\frac{P_1 V_1}{T_1} = \frac{P_2 V_2}{T_2}$$

and, for this problem,

$$V_2 = \frac{P_1 V_1 T_2}{P_2 T_1} = \frac{(732\,\text{torr})(310\,\text{mL})(273\,\text{K})}{(760\,\text{torr})(288\,\text{K})} = 283\,\text{mL}$$

P10.11 The mole fraction is defined in Equation 10.8:

$$X_{O_2} = \frac{P_{O_2}}{P_{total}} = \frac{116\,\text{torr}}{760\,\text{torr}} = 0.153 \text{ or } 15.3\%$$

P10.12 Use Equation 10.10;

$$\frac{\text{effusion rate (HX)}}{\text{effusion rate (HCl)}} = \sqrt{\frac{M_{HCl}}{M_{HX}}}$$

$$M_{HX} = M_{HCl} \times \left(\frac{\text{effusion rate (HX)}}{\text{effusion rate (HCl)}}\right)^2 = 36.46\,\text{g mol}^{-1} \times (1.88)^2 = 128.9\,\text{g mol}^{-1}$$

The unknown gas must be HI.

Review Problems

10.28   (a)   # torr = (1.26 atm)(760 torr / 1 atm) = 958 torr
        (b)   # atm = (740 torr)(1 atm / 760 torr) = 0.974 atm
        (c)   738 torr = 738 mm Hg
        (d)   # torr = (1.45 X 10^3 Pa)(760 torr / 1.01325 X 10^5 Pa) = 10.9 torr

10.30   (a)   # torr = (0.329 atm)(760 torr / 1 atm) = 250 torr
        (b)   # torr = (0.460 atm)(760 torr / 1 atm) = 350 torr

10.32   765 torr − 720 torr = 45 torr   45 torr = 45 mm Hg

$$\text{\# cm Hg} = 45\,\text{mm Hg}\left(\frac{1\,\text{cm}}{10\,\text{mm}}\right) = 4.5\,\text{cm Hg}$$

10.34   65 mm Hg = 65 torr   748 torr + 65 torr = 813 torr

10.36 P_{gas} = P_{atm} + (height difference)
            = 746 mm + 80.0 mm = 826 mm = 826 torr
        When the mercury level in the manometer arm nearest the bulb goes down by 4.00 cm (12.50 to 8.50) the mercury in the other arm goes up by 4.00 cm. Hence, the difference in the heights of the two arms is 8.00 cm = 80.0 mm.

10.38   In a closed-end manometer the difference in height of the mercury levels in the two arms corresponds to the pressure of the gas.

$$\text{\# torr} = (12.5\,\text{cm Hg})\left(\frac{10\,\text{mm}}{1\,\text{cm}}\right)\left(\frac{1\,\text{torr}}{1\,\text{mm Hg}}\right) = 125\,\text{torr}$$

10.40   First convert the pressure in torr to mm Hg:

$$\text{\# mm Hg} = (755\,\text{torr})\left(\frac{1\,\text{mm Hg}}{1\,\text{torr}}\right) = 755\,\text{mm Hg}$$

Then use Equation 10.3 to convert height of the mercury column (liquid A) to height of the fluid column with a density of 1.22 g/mL (liquid B).

$$h_B = h_A \left(\frac{d_A}{d_B} \right) = 755\,\text{mm} \times \frac{13.6\,\text{g mL}^{-1}}{1.22\,\text{g mL}^{-1}} = 8.42 \times 10^3\,\text{mm}$$

10.42   Use Boyle's Law to solve for the second volume:

$$V_2 = \frac{P_1 V_1}{P_2} = \frac{255\,\text{mL}(725\,\text{torr})}{365\,\text{torr}} = 507\,\text{mL}$$

10.44   Use Charles's Law to solve the second volume:

$$V_2 = \frac{V_1 T_2}{T_1} = \frac{3.8\,\text{L}\,(353\,\text{K})}{318\,\text{K}} = 4.28\,\text{L}$$

10.46   Compare pressure change to temperature to solve for temperature change:

$$T_2 = \frac{P_2 T_1}{P_1} = \frac{1700\,\text{torr}(558\,\text{K})}{850\,\text{torr}} = 1120\,\text{K}$$

1120K – 273K = 843 °C

10.48   In general the combined gas law equation is: $\dfrac{P_1 V_1}{T_1} = \dfrac{P_2 V_2}{T_2}$, and in particular, for this problem, we have:

$$P_2 = \frac{P_1 V_1 T_2}{T_1 V_2} = \frac{(740\,\text{torr})(2.58\,\text{L})(348.2\,\text{K})}{(297.2\,\text{K})(2.81\,\text{L})} = 796\,\text{torr}$$

10.50   In general the combined gas law equation is $\dfrac{P_1 V_1}{T_1} = \dfrac{P_2 V_2}{T_2}$, and in particular, for this problem, we have:

$$V_2 = \frac{P_1 V_1 T_2}{T_1 P_2} = \frac{(745\,\text{torr})(2.68\,\text{L})(648.2\,\text{K})}{(297.2\,\text{K})(760\,\text{torr})} = 5.73\,\text{L}$$

10.52   In general the combined gas law equation is: $\dfrac{P_1 V_1}{T_1} = \dfrac{P_2 V_2}{T_2}$, and in particular, for this problem, we have:

$$T_2 = \frac{P_2 V_2 T_1}{P_1 V_1} = \frac{(373\,\text{torr})(9.45\,\text{L})(293.2\,\text{K})}{(761\,\text{torr})(6.18\,\text{L})} = 220\,\text{K} = -53\,^\circ\text{C}$$

10.54 $R = \left(0.0821 \dfrac{\text{L atm}}{\text{mol K}} \right)\left(\dfrac{1000\,\text{mL}}{1\,\text{L}} \right)\left(\dfrac{760\,\text{torr}}{1\,\text{atm}} \right) = 6.24 \times 10^4 \dfrac{\text{mL torr}}{\text{mol K}}$

10.56 $V = \dfrac{nRT}{P} = \dfrac{\left(0.136\text{g}\dfrac{1\,\text{mol}}{32.0\,\text{g}} \right)\left(0.0821 \dfrac{\text{L atm}}{\text{mol K}} \right)(298\,\text{K})}{\left(756\,\text{torr}\dfrac{1\,\text{atm}}{760\,\text{torr}} \right)} = 0.104\,\text{L}$

10.58 $P = \dfrac{nRT}{V} = \dfrac{\left(10.0\text{g}\dfrac{1\,\text{mol}}{32.0\,\text{g}} \right)\left(0.0821 \dfrac{\text{L atm}}{\text{mol K}} \right)(300\,\text{K})}{(2.50\,\text{L})} = 3.08\,\text{atm} \dfrac{760\,\text{torr}}{1\,\text{atm}} = 2343\,\text{torr}$

10.60 $n = \dfrac{PV}{RT} = \dfrac{(0.821\,\text{atm})(0.0265\,\text{L})}{\left(0.0821\,\dfrac{\text{L atm}}{\text{mol K}}\right)(293\,\text{K})} = 9.04 \times 10^{-4}\,\text{mol}\,\dfrac{44\,\text{g}}{1\,\text{mol}} = 0.0398\,\text{g}$

10.62 Since PV = nRT, then;

$$P = \dfrac{nRT}{V} = \dfrac{(4.18\,\text{mol})\left(0.0821\,\frac{\text{L atm}}{\text{mol K}}\right)(291.2\,\text{K})}{(24.0\,\text{L})} = 4.16\,\text{atm}$$

10.64 (a) $\text{density } C_2H_6 = \left(\dfrac{30.1\,\text{g } C_2H_6}{1\,\text{mol } C_2H_6}\right)\left(\dfrac{1\,\text{mol}}{22.4\,\text{L}}\right) = 1.34\,\text{g L}^{-1}$

(b) $\text{density } N_2 = \left(\dfrac{28.0\,\text{g } N_2}{1\,\text{mol } N_2}\right)\left(\dfrac{1\,\text{mol}}{22.4\,\text{L}}\right) = 1.25\,\text{g L}^{-1}$

(c) $\text{density } Cl_2 = \left(\dfrac{70.9\,\text{g } Cl_2}{1\,\text{mol } Cl_2}\right)\left(\dfrac{1\,\text{mol}}{22.4\,\text{L}}\right) = 3.17\,\text{g L}^{-1}$

(d) $\text{density } Ar = \left(\dfrac{39.9\,\text{g Ar}}{1\,\text{mol Ar}}\right)\left(\dfrac{1\,\text{mol}}{22.4\,\text{L}}\right) = 1.78\,\text{g L}^{-1}$

10.66 In general PV = nRT, where n = mass ÷ formula mass. Thus

$$PV = \dfrac{\text{mass}}{(\text{formula mass})} RT$$

and we arrive at the formula for the density (mass divided by volume) of a gas:

$$D = \dfrac{P \times (\text{formula mass})}{RT}$$

$$D = \dfrac{(742\,\text{torr})\left(\frac{1\,\text{atm}}{760\,\text{torr}}\right)(32.0\,\text{g/mol})}{\left(0.0821\,\frac{\text{L atm}}{\text{mol K}}\right)(297.2\,\text{K})}$$

$$D = 1.28\,\text{g/L for } O_2$$

10.68 First determine the number of moles:

$$n = \dfrac{PV}{RT} = \dfrac{(10.0\,\text{torr})\left(\frac{1\,\text{atm}}{760\,\text{torr}}\right)(255\,\text{mL})\left(\frac{1\,\text{L}}{1000\,\text{mL}}\right)}{\left(0.0821\,\frac{\text{L atm}}{\text{mol K}}\right)(298.2\,\text{K})} = 1.37 \times 10^{-4}\,\text{mol}$$

Now calculate the molecular mass:

$$\text{molecular mass} = \dfrac{\text{mass}}{\text{\# of moles}} = \dfrac{(12.1\,\text{mg})\left(\frac{1\,\text{g}}{1000\,\text{mg}}\right)}{1.37 \times 10^{-4}\,\text{mol}} = 88.3\,\text{g/mol}$$

10.70 $\text{molecular mass} = \dfrac{DRT}{P} = \dfrac{\text{mass } RT}{PV} = \dfrac{(1.13\,\text{g/L})\left(0.0821\,\dfrac{\text{L atm}}{\text{mol K}}\right)(295\,\text{K})}{(0.995\,\text{atm})} = 27.5\,\text{g/mol}$

10.72 When gases are held at the same temperature and pressure, and dispensed in this fashion during chemical reactions, then they react in a ratio of volumes that is equal to the ratio of the coefficients (moles) in the balanced chemical equation for the given reaction. We can, therefore, directly use the stoichiometry of the balanced chemical equation to determine the combining ratio of the gas volumes:

$$\# \, L \, F_2 = (4.00 \, L \, H_2)\left(\frac{1 \, \text{volume} \, F_2}{1 \, \text{volume} \, H_2}\right) = 4.00 \, L \, F_2$$

10.74 $$\# \, mL \, O_2 = (175 \, mL \, C_4H_{10})\left(\frac{13 \, mL \, O_2}{2 \, mL \, C_4H_{10}}\right) = 1.14 \times 10^3 \, mL$$

10.76

$$\# \, mol \, C_3H_6 = (18.0 \, g \, C_3H_6)\left(\frac{1 \, mol \, C_3H_6}{42.08 \, g \, C_3H_6}\right) = 0.428 \, mol \, C_3H_6$$

$$\# \, mol \, H_2 = (0.428 \, mol \, C_3H_6)\left(\frac{1 \, mol \, H_2}{1 \, mol \, C_3H_6}\right) = 0.428 \, mol \, H_2$$

$$V = \frac{nRT}{P} = \frac{(0.428 \, mol \, H_2)(0.0821 \frac{L \, atm}{mol \, K})(297.2 \, K)}{(740 \, torr)(\frac{1 \, atm}{760 \, torr})} = 10.7 \, L \, H_2$$

10.78 $CH_4 + 2O_2 \rightarrow CO_2 + 2H_2O$

$$n_{CH_4} = \frac{PV}{RT} = \frac{(725 \, torr)\left(\frac{1 \, atm}{760 \, torr}\right)(16.8 \times 10^{-3} \, L)}{\left(0.0821 \frac{L \, atm}{mol \, K}\right)(308 \, K)} = 6.34 \times 10^{-4} \, \text{moles} \Rightarrow 1.27 \times 10^{-3} \, \text{moles} \, O_2$$

$$V_{O_2} = \frac{nRT}{P} = \frac{(1.27 \times 10^{-3} \, \text{moles})\left(0.0821 \frac{L \, atm}{mol \, K}\right)(300 \, K)}{(654 \, torr)\left(\frac{1 \, atm}{760 \, torr}\right)} = 3.63 \times 10^{-2} \, L = 36.3 \, mL$$

10.80 $2CO + O_2 \rightarrow 2CO_2$

$$\# \, \text{moles} \, CO = \frac{(683 \, torr)\left(\frac{1 \, atm}{760 \, torr}\right)(0.300 \, L)}{\left(0.0821 \frac{L \, atm}{mol \, K}\right)(298 \, K)} = 1.10 \times 10^{-2} \, \text{moles}$$

$$\# \, \text{moles} \, O_2 = \frac{(715 \, torr)\left(\frac{1 \, atm}{760 \, torr}\right)(0.150 \, L)}{\left(0.0821 \frac{L \, atm}{mol \, K}\right)(398 \, K)} \, 4.32 \times 10^{-3} \, \text{moles}$$

$\therefore O_2$ is the limiting reactant

$$\# \, \text{moles} \, CO_2 = (4.32 \times 10^{-3} \, \text{moles} \, O_2)\left(\frac{2 \, mol \, CO_2}{1 \, mol \, O_2}\right) = 8.64 \times 10^{-3} \, \text{moles} \, CO_2$$

$$V = \frac{(8.64 \times 10^{-3} \, mol)\left(0.0821 \frac{L \, atm}{mol \, K}\right)(300 \, K)}{(745 \, torr)\left(\frac{1 \, atm}{760 \, torr}\right)} \, 2.17 \times 10^{-1} \, L = 217 \, mL$$

10.82 $P_{Tot} = 200 \, torr + 150 \, torr + 300 \, torr = 650 \, torr$

10.84   Assume all gases behave ideally and recall that 1 mole of an ideal gas at 0°C and 1 atm occupies a volume of 22.4 L. So,

$$P_{N_2} = 0.30 \text{ atm} = 228 \text{ torr}$$
$$P_{O_2} = 0.20 \text{ atm} = 152 \text{ torr}$$
$$P_{He} = 0.40 \text{ atm} = 304 \text{ torr}$$
$$P_{CO_2} = 0.10 \text{ atm} = 76 \text{ torr}$$

10.86 $P_{total} = (P_{CO} + P_{H_2O})$

$P_{H_2O} = 17.54$ torr at 20 °C, from Table 10.2.

$P_{CO} = 754 - 17.54$ torr $= 736$ torr

The temperature stays constant so, $P_1V_1 = P_2V_2$, and

$$V_2 = \frac{P_1V_1}{P_2} = \frac{(736 \text{ torr})(268 \text{ mL})}{(760 \text{ torr})} = 260 \text{ mL}$$

10.88   From Table 10.2, the vapor pressure of water at 20 °C is 17.54 torr. Thus only $(742 - 17.54) = 724$ torr is due to "dry" methane. In other words, the fraction of the wet methane sample that is pure methane is $724/742 = 0.976$. The question can now be phrased: What volume of wet methane, when multiplied by 0.976, equals 244 mL?

Volume"wet" methane $\times$ 0.976 = 244 mL

Volume"wet" methane = 244 mL/0.976 = 250 mL

In other words, one must collect 250 total mL of "wet methane" gas in order to have collected the equivalent of 244 mL of pure methane.

10.90   Use equation 10.8 to convert all partial pressures to mole fraction. ($P_{total} = 760$ torr)

$$X_{N_2} = \frac{P_{N_2}}{P_{total}} = \frac{570 \text{ torr}}{760 \text{ torr}} = 0.750; 75.0\%$$

$$X_{O_2} = \frac{P_{O_2}}{P_{total}} = \frac{103 \text{ torr}}{760 \text{ torr}} = 0.136; 13.6\%$$

$$X_{CO_2} = \frac{P_{CO_2}}{P_{total}} = \frac{40 \text{ torr}}{760 \text{ torr}} = 0.053; 5.3\%$$

$$X_{H_2O} = \frac{P_{H_2O}}{P_{total}} = \frac{47 \text{ torr}}{760 \text{ torr}} = 0.062; 6.2\%$$

10.92   Effusion rates for gases are inversely proportional to the square root of the gas density, and the gas with the lower density ought to effuse more rapidly. Nitrogen in this problem has the higher effusion rate because it has the lower density:

$$\frac{\text{rate}(N_2)}{\text{rate}(CO_2)} = \sqrt{\frac{1.96 \text{ g L}^{-1}}{1.25 \text{ g L}^{-1}}} = 1.25$$

Additional Exercises

10.96 We found that 1 atm = 33.9 ft of water. This is equivalent to 33.9 ft × 12 in./ft = 407 in. of water, which in this problem is equal to the height of a water column that is uniformly 1.00 in.^{2} in diameter. Next, we convert the given density of water from the units g/mL to the units lb/in.^{3}:

$$\# \frac{lb}{in.^3} = \left(\frac{1.00\,g}{1.00\,mL}\right)\left(\frac{1\,lb}{454\,g}\right)\left(\frac{1\,mL}{1\,cm^3}\right)\left(\frac{2.54\,cm}{1\,in.}\right)^3 = 0.0361\frac{lb}{in.^3}$$

The area of the total column of water is now calculated: 1.00 in.^{2} × 407 in. = 407 in.^{3}, along with the mass of the total column of water: 407 in.^{3} × 0.0361 lb/in.^{3} = 14.7 lb. Finally, we can determine the pressure (force/unit area) that corresponds to one atm: 1 atm = 14.7 lb 1.00 in.^{2} = 14.7 lb/in.^{2}

10.98

$$\text{Total weight} = (45.6\,tons + 8.3\,tons)\left(\frac{2000\,lb}{1\,ton}\right) = 1.08 \times 10^5\,lbs$$

Total pressure = 85 psi + 14.7 psi = 99.7 psi/tire

$$\# \text{ tires} = \frac{1.08 \times 10^5\,lbs}{(99.7\,lbs\,in^{-2}/tire)(100\,in^2)} = 10.8\,tires$$

The minimal number of tires is 12 since tires are mounted in multiples of 2

10.99 Assume a 1 sq in cylinder of water

$$V = (12,468\,ft)\left(\frac{12\,in.}{1\,ft}\right)\left(1\,in.^2\right) = (149616\,in.^3)\left(\frac{2.54\,cm}{1\,in.}\right)^3\left(\frac{1\,mL}{1\,cm^3}\right) = 2.4518 \times 10^6\,mL$$

$$\text{mass} = (2.4518 \times 10^6\,mL)\left(\frac{1.025\,g}{1\,mL}\right)$$

$$= (2.51306 \times 10^6\,g)\left(\frac{1\,lb}{453.6\,g}\right) = 5.54026 \times 10^3\,lb$$

$$\text{Pressure} = (5.54026 \times 10^3\,lb\,in^{-2})\left(\frac{1\,atm}{14.7\,lb\,in^{-2}}\right) = 376.89\,atm$$

10.100 From the data we know that the pressure in flask 1 is greater than atmospheric pressure, and greater than the pressure in flask 2. The pressure in flask 1 can be determined from the manometer data. The pressure in flask 1 is:

$$(0.827\,atm)\left(\frac{760\,mm\,Hg}{1\,atm}\right)\left(\frac{1\,cm}{10\,mm}\right) + 12.26\,cm = 75.11\,cm\,Hg$$

The pressure in flask 2 is lower than flask 1

$$P = 75.11\,cm\,Hg - (16.24\,cm\,oil)\left(\frac{0.826\,g\,mL^{-1}}{13.6\,g\,mL^{-1}}\right) = 74.12\,cm\,Hg = 741.2\,torr$$

10.102 First calculate the initial volume (V_1) and the final volume (V_2) of the cylinder, using the given geometrical data, noting that the radius is half the diameter (10.7/2 = 5.35 cm): $V_1 = \pi \times (5.35\,cm)^2 \times 13.4\,cm = 1.20 \times 10^3\,cm^3$
$V_2 = \pi \times (5.35\,cm)^2 \times (13.4\,cm - 12.7\,cm) = 62.9\,cm^3$

In general the combined gas law equation is: $\dfrac{P_1V_1}{T_1} = \dfrac{P_2V_2}{T_2}$, and in particular, for this problem, we have:

$$T_2 = \frac{P_2V_2T_1}{P_1V_1} = \frac{(34.0\ \text{atm})(62.9\ \text{cm}^3)(364\ \text{K})}{(1.00\ \text{atm})(1.20\ \text{X}\ 10^3\ \text{cm}^3)} = 646\ \text{K} = 373\ ^\circ\text{C}$$

10.104　The temperatures must first be converted to Kelvin:

$$^\circ\text{C} = \frac{5}{9} \times (^\circ\text{F} - 32) = \frac{5}{9} \times (-50 - 32) = -46\ ^\circ\text{C}$$

$$^\circ\text{C} = \frac{5}{9} \times (^\circ\text{F} - 32) = \frac{5}{9} \times (120 - 32) = 49\ ^\circ\text{C}$$

Next, the pressure calculation is done using the following equation:

$$P_2 = \frac{P_1T_2}{T_1} = \frac{(35\ \text{lb in.}^{-2})(322\ \text{K})}{(227\ \text{K})} = 50\ \text{lb in.}^{-2}$$

10.105　$Cl_2 + SO_3^{2-} + H_2O \rightarrow 2Cl^- + SO_4^{2-} + 2H^+$

$$\text{\# moles Cl}_2 = (50.0\ \text{mL Na}_2\text{SO}_3)\left(\frac{0.200\ \text{moles Na}_2\text{SO}_3}{1000\ \text{mL Na}_2\text{SO}_3}\right)\left(\frac{1\ \text{mole SO}_3^{2-}}{1\ \text{mole Na}_2\text{SO}_3}\right)\left(\frac{1\ \text{mole Cl}_2}{1\ \text{mole SO}_3^{\ 2-}}\right)$$

$$= 1.00 \times 10^{-2}\ \text{moles Cl}_2$$

$$V_{Cl_2} = \frac{\left(1.00 \times 10^{-2}\ \text{moles}\right)\left(0.0821\ \dfrac{\text{L atm}}{\text{mol K}}\right)(298\ \text{K})}{(734\ \text{torr})\left(\dfrac{1\ \text{atm}}{760\ \text{torr}}\right)} = 0.253\ \text{L} = 253\ \text{mL}$$

10.108　$P_{total} = 740\ \text{torr} = P_{H_2} + P_{water}$

The vapor pressure of water at 25 °C is available in Table 10.2: 23.76 torr. Hence:
$P_{H_2} = (740 - 24)\ \text{torr} = 716\ \text{torr}$
Next, we calculate the number of moles of hydrogen gas that this represents:

$$n = \frac{PV}{RT} = \frac{(716\ \text{torr})\left(\frac{1\ \text{atm}}{760\ \text{torr}}\right)(0.335\ \text{L})}{\left(0.0821\ \frac{\text{L atm}}{\text{mol K}}\right)(298.2\ \text{K})} = 0.0129\ \text{mol H}_2$$

The balanced chemical equation is: $Zn(s) + 2HCl(aq) \rightarrow H_2(g) + ZnCl_2(aq)$
and the quantities of the reagents that are needed are:

$$\text{\# g Zn} = (0.0129\ \text{mol H}_2)\left(\frac{1\ \text{mol Zn}}{1\ \text{mol H}_2}\right)\left(\frac{65.39\ \text{g Zn}}{1\ \text{mol Zn}}\right) = 0.844\ \text{g Zn}$$

$$\text{\# mL HCl} = (0.0129\ \text{mol H}_2)\left(\frac{2\ \text{mol HCl}}{1\ \text{mol H}_2}\right)\left(\frac{1000\ \text{mL HCl}}{6.00\ \text{mol HCl}}\right) = 4.30\ \text{mL HCl}$$

10.111　We first need to determine the pressure inside the apparatus. Since the water level is 8.5 cm higher inside than outside, the pressure inside the container is lower than the pressure outside. To determine the inside pressure, we first need to convert 8.5 cm of water to an equivalent dimension for mercury. This is done using the density of mercury: $P_{Hg} = 85\ \text{mm}/13.6 = 6.25\ \text{mm}$ (where the density of mercury, 13.6 g/mL, has been used.) $P_{inside} = P_{outside} - P_{Hg} = 764\ \text{torr} - 6\ \text{torr} = 738\ \text{torr}$. In order to determine the P_{H_2}, we need to subtract the vapor pressure of water at 24 °C. This value may be found in Appendix E4 and is equal to 22.4 torr. The $P_{H_2} = P_{inside} - P_{H_2O} =$

738 torr – 22.4 torr = 716 torr. Now, we can use the ideal gas law in order to determine the number of moles of H_2 present;

$$n = \frac{PV}{RT} = \frac{(716 \, \text{torr})\left(\frac{1 \, \text{atm}}{760 \, \text{torr}}\right)(18.45 \, \text{mL})\left(\frac{1 \, L}{1000 \, \text{mL}}\right)}{\left(0.0821 \frac{L \, \text{atm}}{\text{mol K}}\right)(297 \, K)} = 7.13 \times 10^{-4} \, \text{mol}$$

The balanced equation described in this problem is:

$$Zn(s) + 2HCl(aq) \rightarrow ZnCl_2(aq) + H_2(g)$$

By inspection we can see that 1 mole of $Zn(s)$ reacts to form 1 mole of $H_2(g)$ and we must have reacted 7.13×10^{-4} mol Zn in this reaction.

$$\# \, g \, Zn = \left(7.13 \times 10^{-4} \, \text{mol Zn}\right)\left(\frac{65.39 \, g \, Zn}{1 \, \text{mol Zn}}\right) = 4.7 \times 10^{-2} \, g \, Zn$$

Practice Exercises

P11.1   The number of molecules in the vapor will increase, and the number of molecules in the liquid will decrease, but the sum of the molecules in the vapor and the liquid remains the same.

P11.2   We use the curve for water, and find that at 330 torr, the boiling point is approximately 75 °C.

P11.3   Using Equation 11.2 and data from Example 11.1 we calculate;

$$\ln \frac{P_1}{P_2} = \frac{\Delta H_{vap}}{R} \left(\frac{1}{T_2} - \frac{1}{T_1} \right)$$
$$= \frac{30.1 \times 10^3 \, J \, mol^{-1}}{8.314 \, J \, mol^{-1} \, K^{-1}} \left(\frac{1}{333 \, K} - \frac{1}{298 \, K} \right)$$
$$= \left(3.62 \times 10^3 \, K \right) \times \left(-3.53 \times 10^{-4} \, K^{-1} \right)$$
$$= -1.28$$

Taking the antilogarithm we get

$$\frac{P_1}{P_2} = e^{-1.28} = 0.278$$

$$P_2 = \frac{P_1}{0.278} = \frac{148 \, torr}{0.278} = 532 \, torr$$

P11.4   Adding heat will shift the equilibrium to the right, producing more vapor. This increase in the amount of vapor causes a corresponding increase in the pressure, such that the vapor pressure generally increases with increasing temperature.

P11.5   Refer to the phase diagram for water, Figure 11.24. We "move" along a horizontal line marked for a pressure of 2.15 torr. At –20 °C, the sample is a solid. If we bring the temperature from –20 °C to 50 °C, keeping the pressure constant at 2.15 torr, the sample becomes a gas. The process is thus solid → gas, i.e. sublimation.

P11.6   As diagramed in Figure 11.24, this falls in the liquid region.

P11.7   Because this is a high melting, hard material, it must be a covalent or network solid. Covalent bonds link the various atoms of the crystal.

P11.8   Since the melt does not conduct electricity, it is not an ionic substance. The softness and the low melting point suggest that this is a molecular solid, and indeed the formula is most properly written S_8.

Review Problems

11.91   Diethyl ether has the faster rate of evaporation, since it does not have hydrogen bonds, as does butanol.

11.93   London forces are possible in them all. Where another intermolecular force can operate, it is generally stronger than London forces, and this other type of interaction overshadows the importance of the London force. The substances in the list that can have dipole–dipole attractions are those with permanent dipole moments: (a), (b), and (d). SF_6, (c), is a non–polar molecular substance. HF, (a), has hydrogen bonding.

11.95   Ethanol, because it has H-bonding.

11.97   ether < acetone < benzene < water < acetic acid

11.99 $\# kJ = (125 \, g \, H_2O)\left(\dfrac{1 \, mol \, H_2O}{18.015 \, g \, H_2O}\right)\left(\dfrac{43.9 \, kJ}{1 \, mol \, H_2O}\right) = 305 \, kJ$

11.101 We can approach this problem by first asking either of two equivalent questions about the system: how much heat energy (q) is needed in order to melt the entire sample of solid water (105 g), or how much energy is lost when the liquid water (45.0 g) is cooled to the freezing point? Regardless, there is only one final temperature for the combined (150.0 g) sample, and we need to know if this temperature is at the melting point (0 °C, at which temperature some solid water remains in equilibrium with a certain amount of liquid water) or above the melting point (at which temperature all of the solid water will have melted).

Heat flow supposing that all of the solid water is melted:
$\quad q = 6.01 \, kJ/mole \times 105 \, g \times 1 \, mol/18.0 \, g = 35.1 \, kJ$

Heat flow on cooling the liquid water to the freezing point:
$\quad q = 45.0 \, g \times 4.18 \, J/g \, °C \times 85 \, °C = 1.60 \times 10^4 \, J = 16.0 \, kJ$

The lesser of these two values is the correct one, and we conclude that 16.0 kJ of heat energy will be transferred from the liquid to the solid, and that the final temperature of the mixture will be 0 °C. The system will be an equilibrium mixture weighing 150 g and having some solid and some liquid in equilibrium with one another. The amount of solid that must melt in order to decrease the temperature of 45.0 g of water from 85 °C to 0 °C is: 16.0 kJ ÷ 6.01 kJ/mol = 2.66 mol of solid water. 2.66 mol × 18.0 g/mol = 47.9 g of water must melt.

(a)   The final temperature will be 0 °C.
(b)   47.9 g of water must melt.

11.103

$$\ln\left(\dfrac{P_1}{P_2}\right) = \dfrac{\Delta H_{vap}}{8.314 \, J \, mol^{-1}K^{-1}}\left(\dfrac{1}{T_2} - \dfrac{1}{T_1}\right)$$

$$\ln\left(\dfrac{72.8}{186.2}\right) = \dfrac{\Delta H_{vap}}{8.314 \, J \, mol^{-1}K^{-1}}\left(\dfrac{1}{313 \, K} - \dfrac{1}{293 \, K}\right)$$

$$-0.939 \times 8.314 \, J \, mol^{-1}K^{-1} = \Delta H_{vap} \times \left(-0.000218 \, K^{-1}\right)$$

$$\Delta H_{vap} = 3.58 \times 10^4 \, J \, mol^{-1} = 35.8 \, kJ/mol$$

11.105

$$\ln\left(\dfrac{P_1}{P_2}\right) = \dfrac{\Delta H_{vap}}{8.314 \, J \, mol^{-1}K^{-1}}\left(\dfrac{1}{T_2} - \dfrac{1}{T_1}\right)$$

$$\ln\left(\dfrac{28.1}{47.0}\right) = \dfrac{\Delta H_{vap}}{8.314 \, J \, mol^{-1}K^{-1}}\left(\dfrac{1}{60 \, K} - \dfrac{1}{57 \, K}\right)$$

$$-0.514 \times 8.314 \, J \, mol^{-1}K^{-1} = \Delta H_{vap} \times \left(-8.77 \times 10^{-4} \, K^{-1}\right)$$

$$\Delta H_{vap} = 4.87 \times 10^3 \, J \, mol^{-1} = 4.87 \, kJ/mol$$

The normal boiling point is where the vapor pressure equals 760 torr. So,

$$\ln\left(\frac{P_1}{P_2}\right) = \frac{\Delta H_{vap}}{8.314\,J\,mol^{-1}K^{-1}}\left(\frac{1}{T_2} - \frac{1}{T_1}\right)$$

$$\ln\left(\frac{28.1}{760}\right) = \frac{\Delta H_{vap}}{8.314\,J\,mol^{-1}K^{-1}}\left(\frac{1}{T_2} - \frac{1}{57\,K}\right)$$

$$-3.298 \times 8.314\,J\,mol^{-1}K^{-1}/\,4.87 \times 10^3\,J/mol = \frac{1}{T_2} - \frac{1}{57\,K}$$

$$-5.64 \times 10^{-3}\,K^{-1} = \frac{1}{T_2} - 1.75 \times 10^{-2}$$

$$\frac{1}{T_2} = 1.19 \times 10^{-2}$$

$$T_2 = 84.0\,K = -189.2\,^\circ C$$

11.107

$$\ln\left(\frac{P_1}{P_2}\right) = \frac{\Delta H_{vap}}{8.314\,J\,mol^{-1}K^{-1}}\left(\frac{1}{T_2} - \frac{1}{T_1}\right)$$

$$\ln\left(\frac{185}{P_2}\right) = \frac{38.0 \times 10^3\,Jmol^{-1}}{8.314\,J\,mol^{-1}K^{-1}}\left(\frac{1}{298\,K} - \frac{1}{291.4\,K}\right)$$

$$= -0.347$$

$$P_2 = \frac{20.0\,torr}{\exp(-0.347)} = 28.3\,torr$$

11.109

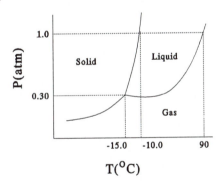

11.111 (a) solid (b) gas (c) liquid (d) solid, liquid, and gas

11.113 At −56 °C, the vapor is compressed until the liquid–vapor line is reached, at which point the vapor condenses to a liquid. As the pressure is increased further, the solid–liquid line is reached, and the liquid freezes. At −58°C, the gas is compressed until the solid–vapor line is reached, at which point the vapor condenses directly to a solid.

11.115 Each atom of a cubic unit cell is shared by eight unit cells. This means that only one eighth of each corner atom can be assigned to a given unit cell. Eight corner atoms times 1/8 each assigned to a given unit cell yields one atom per unit cell.

11.117 In the following diagram, the dashed line is a face diagonal of the unit cell. The unit cell edge is 362 pm. By the Pythagorean theorem, we have: diagonal^2 = edge^2 + edge^2, where an edge is equal to 362 pm. Also, as shown in the diagram below, the face diagonal is equal to 4 × r, where r equals the atom's radius.

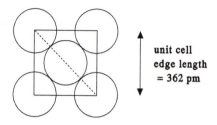

unit cell
edge length
= 362 pm

This means that $[4r]^2 = (362)^2 + (362)^2$. Taking the square root of both sides of this equation gives: $4r = 1.414 \times 362$ pm, and we get $r = 128$ pm.

11.119 Each edge is composed of 2 × radius of the cation plus 2 × radius of the anion. The edge is therefore 2 × 133 + 2 × 195 = 656 pm.

11.121 This must be a molecular solid, because if it were ionic it would be high–melting, and the melt would conduct.

11.123 This is a metallic solid.

11.125 This is metallic.

<u>Practice Exercises</u>

P12.1 Since these solutions are saturated, the maximum amount of each gas is dissolved in the solution. Hence, 0.00430 g of O_2 and 0.00190 g of N_2 are dissolved in the water.

P12.2 The total mass of the solution is to be 250 g. If the solution is to be 1.00 % (w/w) NaOH, then the mass of NaOH will be: 250 g × 1.00 g NaOH/100 g solution = 2.50 g NaOH. We therefore need 2.50 g of NaOH and (250 – 2.50) = 248 g H_2O. The volume of water that is needed is: 248 g ÷ 0.988 g/mL = 251 mL.

P12.3 An HCl solution that is 37 % (w/w) has 37 grams of HCl for every 1.0×10^2 grams of solution.

$$\text{\# g solution} = \left(7.5\,\text{g HCl}\right)\left(\frac{1.0 \times 10^2\,\text{g solution}}{37\,\text{g HCl}}\right) = 2.0 \times 10^1\,\text{g solution}$$

P12.4
$$\text{\# g CH}_3\text{OH} = \left(2000\,\text{g H}_2\text{O}\right)\left(\frac{0.250\,\text{mol CH}_3\text{OH}}{1000\,\text{g H}_2\text{O}}\right)\left(\frac{32.0\,\text{g CH}_3\text{OH}}{1\,\text{mol CH}_3\text{OH}}\right)$$
$$= 16.0\,\text{g CH}_3\text{OH}$$

P12.5 We need to know the number of moles of NaOH and the number of kg of water.
4.00 g NaOH ÷ 40.0 g/mol = 0.100 mol NaOH
250 g H_2O × 1 kg/1000 g = 0.250 kg H_2O

The molality is thus given by:
m = 0.100 mol/0.250 kg = 0.400 mol NaOH/kg H_2O

P12.6 If a solution is 37.0 % (w/w) HCl, then 37.0 % of the mass of any sample of such a solution is HCl and (100.0 – 37.0) = 63.0 % of the mass is water. In order to determine the molality of the solution, we can conveniently choose 100.0 g of the solution as a starting point. Then 37.0 g of this solution are HCl and 63.0 g are H_2O. For molality, we need to know the number of moles of HCl and the mass in kg of the solvent:
37.0 g HCl ÷ 36.46 g/mol = 1.01 mol HCl
63.0 g H_2O × 1 kg/1000 g = 0.0630 kg H_2O
molality = mol HCl/kg H_2O = 1.01 mol/0.0630 kg = 16.1 m

P12.7 We first determine the mass of one L (1000 mL) of this solution, using the density:
1000 mL × 1.38 g/mL = 1.38×10^3 g
Next, we use the fact that 40.0 % of this total mass is due to HBr, and calculate the mass of HBr in the 1000 mL of solution: 0.400 × 1.38×10^3 = 552 g HBr.
This is converted to the number of moles of HBr in 552 g: 552 g HBr ÷ 80.91 g/mol = 6.82 mol HBr. Last, the molarity is the number of moles of HBr per liter of solution: 6.82 mol/1 L = 6.82 M

P12.8 First determine the number of moles of each component of the solution:
For $C_{16}H_{22}O_4$, 20.0 g/278 g/mol = 0.0719 mol
For C_8H_{18}, 50.0 g/114 g/mol = 0.439 mol
and the mole fraction of solvent is: 0.439 mol/(0.439 mol + 0.0719 mol) = 0.859
Using Raoult's Law, we next find the vapor pressure to expect for the solution, which arises only from the solvent (since the solute is known to be nonvolatile):

$$P_{solvent} = X_{solvent} \times P°_{solvent} = 0.859 \times 10.5\,\text{torr} = 9.02\,\text{torr}$$

P12.9 $P_{cyclohexane} = X_{cyclohexane} \times P°_{cyclohexane} = 0.500 \times 66.9\,\text{torr} = 33.4\,\text{torr}$
$P_{toluene} = X_{toluene} \times P°_{toluene} = 0.500 \times 21.1\,\text{torr} = 10.6\,\text{torr}$
$P_{total} = P_{cyclohexane} + P_{toluene} = 33.4\,\text{torr} + 10.6\,\text{torr} = 44.0\,\text{torr}$

P12.10 A 10 % solution contains 10 g sugar and 90 g water.

10 g $C_{12}H_{22}O_{11}$ ÷ 342 g/mol = 0.029 mol $C_{12}H_{22}O_{11}$

90 g H_2O × 1 kg/1000 g = 0.090 kg H_2O

m = 0.029 mol/0.090 kg = 0.32 mol/kg

$\Delta T_b = K_b \times m$ = 0.51 °C m^{-1} × 0.32 m = 0.16 °C

T_b = 100.16 °C

P12.11 It is first necessary to obtain the values of the freezing point of pure benzene and the value of K_f for benzene from Table 12.4 of the text. We proceed to determine the number of moles of solute that are present and that have caused this depression in the freezing point:

$\Delta T = K_f m$, ∴ $m = \Delta T/K_f$ = (5.45 °C – 4.13 °C)/(5.07 °C kg mol^{-1}) = 0.260 m

Next, use this molality to determine the number of moles of solute that must be present:

0.260 mol solute/kg solvent × 0.0850 kg solvent = 0.0221 mol solute

Last, determine the formula weight of the solute: 3.46 g/0.0221 mol = 157 g/mol

P12.12 Π = MRT

Π = (0.0115 mol/L)(0.0821 L atm/K mol)(298 K) = 0.281 atm

0.281 atm × 760 torr/atm = 214 torr

P12.13 We can use the equation Π = MRT, remembering to convert pressure to atm:

$$\# \, atm = (25.0 \, torr)\left(\frac{1 \, atm}{760 \, torr}\right) = 0.0329 \, atm$$

Π = 0.0329 atm = M × (0.0821 L atm/K mol)(298 K)

M = 1.34 X 10^{-3} mol L^{-1}

mol = 1.34 X 10^{-3} mol L^{-1} × 0.100 L = 1.34 X 10^{-4} mol

$$formula \; mass = \frac{72.4 \, X \, 10^{-3} \, g}{1.34 \, X \, 10^{-4} \, mol} = 5.38 \, X \, 10^2 \, g \, mol^{-1}$$

P12.14 For the solution as if the solute were 100 % dissociated:

ΔT = (1.86 °C m^{-1})(2 × 0.237 m) = 0.882 °C and the freezing point should be –0.882 °C.

For the solution as if the solute were 0 % dissociated:

ΔT = (1.86 °C m^{-1})(1 × 0.237 m) = 0.441 °C and the freezing point should be –0.441 °C.

Review Problems

12.57 This is to be very much like that shown in Figures 12.7 and 12.8:

(a)	KCl(s) → K^+(g) + Cl^-(g),	$\Delta H°$ = +690 kJ mol^{-1}
(b)	K^+(g) + Cl^-(g) K^+(aq) + Cl^-(aq),	$\Delta H°$ = –686 kJ mol^{-1}
	KCl(s) K^+(aq) + Cl^-(aq),	$\Delta H°$ = +4 kJ mol^{-1}

12.59 $C_1/P_1 = C_2/P_2$, $C_2 = (C_1 \times P_2)/P_1$ = (0.025 g/L × 1.5 atm)/1.0 atm = 0.038 g/L.

12.61 24.0 g glucose ÷ 180 g/mol = 0.133 mol glucose

molality = 0.133 mol glucose/1.00 kg solvent = 0.133 molal

mole fraction = moles glucose/total moles

moles glucose = 0.133

$$\text{moles H}_2\text{O} = (1.00 \times 10^3 \text{ g H}_2\text{O})\left(\frac{1\,\text{mole H}_2\text{O}}{18\,\text{g H}_2\text{O}}\right) = 55.5 \text{ moles H}_2\text{O}$$

$$\text{Xglucose} = \frac{1.33}{55.5 + 0.133} = 2.39 \times 10^{-3}$$

weight % = 24.0/1024 x 100% = 2.34%

12.63   We need to know the mole amounts of both components of the mixture. It is convenient to work from an amount of solution that contains 1.25 mol of ethyl alcohol and, therefore, 1.00 kg of solvent. Convert the number of moles into mass amounts as follows:

For CH_3CH_2OH, 1.25 mol × 46.1 g/mol = 57.6 g

Thus, the total mass is (57.6 + 1000) g = 1058 g, and the mass % of alcohol is given by: 57.6 g/1058 g × 100 = 5.44 %

12.65   If we assume 100g of solution we have 5 g NH_3 and 95 g H_2O.

$$\text{\# moles NH}_3 = \left(5.00 \text{ g NH}_3\right)\left(\frac{1\,\text{mole NH}_3}{17.03\,\text{g NH}_3}\right) = 0.294 \text{ moles NH}_3$$

$$\text{\# kg H}_2\text{O} = \left(95.0 \text{ g}\right)\left(\frac{1\,\text{kg}}{1000\,\text{g}}\right) = 0.095 \text{ kg water}$$

$$m = \frac{0.294 \text{ moles}}{0.095 \text{ kg}} = 3.09 \text{ m}$$

$$\text{\# moles H}_2\text{O} = \left(95.0 \text{ g}\right)\left(\frac{1\,\text{mole H}_2\text{O}}{18\,\text{g H}_2\text{O}}\right) = 5.27 \text{ moles H}_2\text{O}$$

mole percent = 0.294 moles/(5.27 moles + 0.294 moles) x 100% = 5.28%

12.67   If we choose, for convenience, an amount of solution that contains 1 kg of solvent, then it also contains 0.363 moles of $NaNO_3$. The number of moles of solvent is:

1.00×10^3 g ÷ 18.0 g/mol = 55.6 mol H_2O

Now, convert the number of moles to a number of grams: for $NaNO_3$, 0.363 mol × 85.0 g/mol = 30.9 g; for H_2O, 1000 g was assumed and the percent (w/w) values are:

% $NaNO_3$ = 30.9 g/1031 g × 100 = 3.00 %

% H_2O = 1000 g/1031 g × 100 = 97.0 %

Now determine the mass of 1.00 L of this solution, using the known density: 1000 mL × 1.0185 g/mL = 1018.5 g. Next, we use the fact that 3.00 % of any sample of this solution is $NaNO_3$, and calculate the mass of $NaNO_3$ that is contained in 1000 mL of the solution: 0.0300 × 1018.5 g = 30.6 g $NaNO_3$. The number of moles of $NaNO_3$ is given by: 30.6 g ÷ 85.0 g/mol = 0.359 mol $NaNO_3$. The molarity is the number of moles of $NaNO_3$ per liter of solution: 0.359 mol/1.00 L = 0.359 M. Once again assume1 kg of solvent so we have 0.363 moles $NaNO_3$.

$$\text{\# mol H}_2\text{O} = \left(1000 \text{ g}\right)\left(\frac{1\,\text{mol}}{18.02\,\text{g}}\right) = 55.6 \text{ moles H}_2\text{O}$$

$$X_{NaNO_3} = \frac{0.363}{55.6 + 0.363} = 6.49 \times 10^{-3}$$

12.69 $P_{solution} = P^°_{solvent} \times X_{solvent}$
We need to determine $X_{solvent}$:

$$\# \text{ mol glucose} = (65.0 \text{ g})\left(\frac{1 \text{ mol}}{180.2 \text{ g}}\right) = 0.361 \text{ moles}$$

$$\# \text{ mol } H_2O = (150 \text{ g } H_2O)\left(\frac{1 \text{ mol } H_2O}{18.02 \text{ g } H_2O}\right) = 8.32 \text{ mol } H_2O$$

The total number of moles is thus: 8.32 mol + 0.361 mol = 8.69 mol and the mole fraction of the solvent is:

$$X_{solvent} = \left(\frac{8.32 \text{ mol solvent}}{8.69 \text{ mol solution}}\right) = 0.957. \text{ Therefore,}$$

$P_{solution}$ = 23.8 torr × 0.957 = 22.8 torr.

12.71 $P_{benzene} = X_{benzene} \times P^°_{benzene}$
$P_{toluene} = X_{toluene} \times P^°_{toluene}$
$P_{Tot} = P_{benzene} + P_{Toluene}$

$$\# \text{ mol benzene} = (60.0 \text{ g})\left(\frac{1 \text{ mol}}{78.11 \text{ g}}\right) = 0.768 \text{ mol benzene}$$

$$\# \text{ mol toluene} = (40.0 \text{ g})\left(\frac{1 \text{ mol}}{92.14 \text{ g}}\right) 0.434 \text{ mol toluene}$$

$$X_{benzene} = \frac{0.768}{0.768 + 0.434} = 0.639 \ .$$

$$X_{toluene} = \frac{0.434}{0.768 + 0.434} = 0.361$$

$P_{benzene} = (0.639)(93.4 \text{ torr}) = 59.7 \text{ torr}$
$P_{toluene} = (0361)(26.9 \text{ torr}) = 9.71 \text{ torr}$
$P_{Tot} = 59.7 \text{ torr} + 9.71 \text{ torr} = 69.4 \text{ torr}$

12.73   The following relationships are to be established: P_{Total} = 96 torr = $P^°_{benzene} \times X_{benzene} + P^°_{Toluene} \times X_{Toluene}$. The relationship between the two mole fractions is: $X_{benzene} = 1 - X_{Toluene}$, since the sum of the two mole fractions is one. Substituting this expression for $X_{benzene}$ into the first equation gives: 96 torr = $P^°_{benzene} \times (1 - X_{Toluene})$ + $P^°_{Toluene} \times X_{Toluene}$, 96 torr = 180 torr × $(1 - X_{Toluene})$ + 60 torr × $X_{Toluene}$. Solving for $X_{Toluene}$ we get: 120 × $X_{Toluene}$ = 84, $X_{Toluene}$ = 0.70 and $X_{benzene}$ = 0.30. The mole % values are to be 70 % toluene and 30 % benzene.

12.75   (a) $X_{solvent} = P/P^° = 511 \text{ torr}/526 \text{ torr} = 0.971$
         $X_{solute} = 1 - X_{solvent} = 0.029$
         (b)   We know 0.971 = 1 mol/1 mol + x moles   x = 2.99×10^{-2} moles
         (c)   molar mass = 8.3 g/2.99×10^{-2} moles = 278 g/mol

12.77 $\Delta T = K_b m = 0.51 \text{ °C kg mol}^{-1} \times 2.0 \text{ mol kg}^{-1} = 1.0 \text{ °C}$
         ∴ T_b = 100.0 + 1.0 = 101 °C

         $\Delta T = K_f m = 1.86 \text{ °C kg mol}^{-1} \times 2.00 \text{ mol kg}^{-1} = 3.72 \text{ °C}$
         ∴ T_f = 0.0 − 3.72 = −3.72 °C

12.79 $\Delta T_f = K_f m$

$m = \Delta T_f / K_f = 3.00\ °C / 1.86\ °C\ kg/mol = 1.61\ mol/kg$

$$\#\ kg = (100\ g)\left(\frac{1\ kg}{1000\ g}\right) = 0.1\ kg$$

$$\#\ mol = (1.61\ mol/kg)(0.1\ kg) = 0.161\ mol$$

$$\#\ g = (0.161\ mol)\left(\frac{342.3\ g}{1\ mol}\right) = 55.1\ g$$

12.81 $\Delta T = (5.45 - 3.45) = 2.00\ °C = K_f \times m = 5.07\ °C\ kg\ mol^{-1} \times m$

∴ $m = 0.394\ mol\ solute/kg\ solvent$

$0.394\ mol/kg\ benzene \times 0.200\ kg\ benzene = 0.0788\ mol\ solute$ and the molecular weight is: $12.00\ g/0.0788$ mol = 152 g/mol

12.83 $\Delta T_f = K_f m$

$m = \Delta T_f / K_f = 0.307\ °C / 5.07\ °C\ kg/mol = 0.0606\ mol/kg$

$\#\ mol = (0.0606\ mol/kg)(0.5\ kg) = 0.0303\ mol$

$$molar\ mass = \frac{3.84\ g}{0.0303\ mol} = 127\ g/mol$$

The empirical formula has a mass of 64.1 g/mol. So the molecular formula is

$C_8 H_4 N_2$

12.85   (a)   If the equation is correct, the units on both sides of the equation should be g/mol. The units on the right side of this equation are:

$$\frac{(g) \times (L\ atm\ mol^{-1}\ K^{-1}) \times (K)}{L \times atm} = g/mol$$

which is correct.

(b) $\Pi = MRT = (n/V)RT,\ \ n = \Pi V/RT$

This means that we can calculate the number of moles of solute in one L of solution, as follows:

$$n = \frac{(0.021\ torr)\left(1\ atm/760\ torr\right)(1.0\ L)}{(0.0821\ L\ atm\ mol^{-1}\ K^{-1})(298\ K)} = 1.1 \times 10^{-6}\ mol$$

The molecular mass is the mass in 1 L divided by the number of moles in 1 L:

$2.0\ g / 1.1 \times 10^{-6}\ mol = 1.8 \times 10^{6}\ g/mol$

12.87   The equation for the vapor pressure is:

$P_{solution} = P°_{H_2O} \times X_{H_2O}$

Where $P°_{H_2O}$ is 17.5 torr. To calculate the vapor pressure we need to find the mole fraction of water first.

$X_{H_2O} = moles\ H_2O / (moles\ H_2O + moles\ NaCl)$

Calculate the moles of NaCl in 10.0 g

$$\#\ mol\ NaCl = (10.0\ g\ NaCl)\left(\frac{1\ mol\ NaCl}{58.44\ g\ NaCl}\right) = 0.171\ moles\ NaCl$$

When NaCl dissolves in water, Na^+ and Cl^- are formed. So, for every mole of NaCl that dissolves, two moles of ions are formed. For this solution, the number of moles of ions is 0.342.

The number of moles of solvent (water) is:

$$\#\ mol\ H_2O = (100\ g\ H_2O)\left(\frac{1\ mol\ H_2O}{18.02\ g\ H_2O}\right) = 5.55\ moles\ H_2O$$

Calculate the mole fraction as

$$X_{H_2O} = \frac{(\text{moles } H_2O)}{(\text{moles } H_2O + \text{moles } NaCl)} = \frac{5.55 \, mol}{(5.55 \, mol + 0.342 \, mol)} = 0.942$$

The vapor pressure is then $P_{solution} = P°_{H_2O} \times X_{H_2O} = 17.5 \, torr \times 0.942 = 16.5 \, torr$

12.89 $\Pi = MRT$

$$M = \frac{(2.0 \, g \, NaCl)\left(\dfrac{1 \, mol \, NaCl}{58.45 \, g \, NaCl}\right)}{0.100 \, L} = 0.34 \, M$$

For every NaCl there are two ions produced so $M = 0.68 \, M$

$$\Pi = (0.68 \, M)(0.0821 \, Latm/molK)(298 \, K)\left(\frac{760 \, torr}{1 \, atm}\right) = 1.3 \times 10^3 \, torr$$

12.91 $CaCl_2 \rightarrow Ca^{2+} + 2Cl^-$; van't Hoff factor, $i = 3$
$\Delta T_f = i \times k_f \times m = (3)(1.86 \, °C \, m^{-1})(0.20 \, m) = 1.1 \, °C$
The freezing point is $-1.1 \, °C$.

12.93   The freezing point depression that is expected from this solution if HF behaves as a nonelectrolyte is: $\Delta T_f = 1.86 \, °C/m \times 1.00 \, m = 1.86 \, °C$. The freezing point that is expected upon complete dissociation of HF is: $\Delta T_f = K_f \times (2 \times m) = 3.72 \, °C$. The observed freezing point depression is 1.91 °C, and the apparent molality is: $m = \Delta T_f/K_f = (1.91 \, °C)/(1.86 \, °C/m) = 1.03$ mol solute particles per kg of solvent. This represents a mole excess of 3 solute particles per mol of HF ($1.03 \, m - 1.00 \, m = 0.03 \, m$)., and we conclude that the percent ionization is 3 %.

12.95   Any electrolyte such as $NiSO_4$, that dissociated to give 2 ions, if fully dissociated should have a van't Hoff factor of 2.

12.97 $\Delta T_f = i \times k_f \times m$

$i = \Delta T_f/k_f \times m = 0.415°C/(1.86 \, °Ccm^{1-})(0.118 \, m) = 1.89$

Additional Problems

12.99   Solubility = 0.015 g/L

$$\frac{\# \, moles}{L} = \left(\frac{0.015 \, g}{L}\right)\left(\frac{1 \, mol}{28.01 \, g}\right) = 5.4 \times 10^{-4} \, mol$$

At 100 feet P = 4 atm. Using Henry's Law $\dfrac{C_1}{P_1} = \dfrac{C_2}{P_2}$ we can determine the solubility of $N_2 = 0.060 g/L$

$$\frac{\# \, moles}{L} = \left(\frac{0.060 \, g}{L}\right)\left(\frac{1 \, mol}{28.01 \, g}\right) = 2.1 \times 10^{-3} \, mol$$

The difference is 1.6×10^{-3} moles

$$V = n/PRT = \frac{(1.6 \times 10^{-3} \, moles)(0.0821 \, \frac{L \, atm}{mol \, K})(310 \, K)}{1 \, atm}$$
$$= 4.07 \times 10^{-2} \, L = 41 \, mL$$

12.101 (a) The formula weights are $Na_2Cr_2O_7 \cdot 2H_2O$: 298 g/mol, C_3H_8O: 60.1 g/mol, and C_3H_6O: 58.1 g/mol.

$$\# \text{ g } Na_2Cr_2O_7 \cdot 2H_2O = (21.4 \text{ g } C_3H_8O)\left(\frac{1 \text{ mol } C_3H_8O}{60.1 \text{ g } C_3H_8O}\right)$$

$$\times \left(\frac{1 \text{ mol } Na_2Cr_2O_7 \cdot 2H_2O}{3 \text{ mol } C_3H_8O}\right)\left(\frac{298 \text{ g } Na_2Cr_2O_7 \cdot 2H_2O}{1 \text{ mol } Na_2Cr_2O_7 \cdot 2H_2O}\right)$$

$$= 35.4 \text{ g } Na_2Cr_2O_7 \cdot 2H_2O$$

(b) The theoretical yield is:

$$\# \text{ g } C_3H_6O = (21.4 \text{ g } C_3H_8O)\left(\frac{1 \text{ mol } C_3H_8O}{60.1 \text{ g } C_3H_8O}\right)\left(\frac{3 \text{ mol } C_3H_6O}{3 \text{ mol } C_3H_8O}\right)\left(\frac{58.1 \text{ g } C_3H_6O}{1 \text{ mol } C_3H_6O}\right)$$

$$= 20.7 \text{ g } C_3H_6O$$

The percent yield is therefore: $12.4/20.7 \times 100 = 59.9 \%$

(c) First, we determine the number of grams of C, H, and O that are found in the products, and then the % by weight of C, H, and O that were present in the sample that was analyzed by combustion, i.e. the by–product:

$$\# \text{ g } C = (22.368 \times 10^{-3} \text{ g } CO_2)\left(\frac{12.011 \text{ g } C}{44.010 \text{ g } CO_2}\right) = 6.1046 \times 10^{-3} \text{ g } C$$

and the % C is: $6.1046 \times 10^{-3} \text{ g}/8.654 \times 10^{-3} \text{ g} \times 100 = 70.54 \% \text{ C}$

$$\# \text{ g } H = (10.655 \times 10^{-3} \text{ g } H_2O)\left(\frac{2.0159 \text{ g } H}{18.015 \text{ g } H_2O}\right) = 1.1923 \times 10^{-3} \text{ g } H$$

and the % H is: $1.1923 \times 10^{-3} \text{ g } H/8.654 \times 10^{-3} \text{ g} \times 100 = 13.78 \% \text{ H}$

For O, the mass is the total mass minus that of C and H in the sample that was analyzed:

8.654×10^{-3} g total – $(6.1046 \times 10^{-3}$ g C + 1.1923×10^{-3} g H) = 1.357×10^{-3} g O
and the % O is: 1.357×10^{-3} g$/8.654 \times 10^{-3}$ g $\times 100 = 15.68 \%$ O. Alternatively, we could have determined the amount of oxygen by using the mass % values, realizing that the sum of the weight percent values should be 100. Next, we convert these mass amounts for C, H, and O into mole amounts by dividing the amount of each element by the atomic weight of each element:
For C, 6.1046×10^{-3} g C $\div 12.011$ g/mol = 0.50825×10^{-3} mol C
For H, 1.1923×10^{-3} g H $\div 1.0079$ g/mol = 1.1829×10^{-3} mol H
For O, 1.357×10^{-3} g O $\div 16.00$ g/mol = 0.08481×10^{-3} mol O
Lastly, these are converted to relative mole amounts by dividing each of the above mole amounts by the smallest of the three (We can ignore the 10^{-3} term since it is common to all three components):
For C, 0.50825 mol/0.08481 mol = 5.993
For H, 1.1829 mol/0.08481 mol = 13.95
For O, 0.08481 mol/0.08481 mol = 1.000
and the empirical formula is given by this ratio of relative mole amounts, namely $C_6H_{14}O$.

(d) $\Delta T_f = K_f m$, (5.45 °C – 4.87 °C) = (5.07 °C/m) $\times m$, $m = 0.11$ molal, and there are 0.11 moles of solute dissolved in each kg of solvent. Thus, the number of moles of solute that have been used here is:
0.11 mol/kg $\times$ 0.1150 kg = 1.31×10^{-2} mol solute.
The formula weight is thus: 1.338 g/0.0131 mol = 102 g/mol. Since the empirical formula has this same mass, we conclude that the molecular formula is the same as the empirical formula, i.e. $C_6H_{14}O$.

12.104 (a) The height difference is proportional to the osmotic pressure, therefore Π may be calculated by converting the height difference to the height of a mercury column in mm, which is equal to the pressure in torr (1 mm Hg = 1 torr):

$$h_{Hg} = h_{solution} \times (d_{solution}/d_{Hg}) = (12.6 \text{ mm}) \times (1.00 \text{ g/mL}/13.6 \text{ g/mL}) = 0.926 \text{ mm Hg}$$
$$P = 0.926 \text{ torr}$$

(b) $\Pi = MRT$

$$M = \Pi/RT = \frac{(0.926 \text{ torr})\left(\dfrac{1 \text{ atm}}{760 \text{ torr}}\right)}{\left(0.0821 \dfrac{L \text{ atm}}{\text{mol K}}\right)(298 \text{ K})} = 4.98 \times 10^{-5} \text{ M}$$

(c) Since this is a dilute solution and the solute does not dissociate, we can assume that the molarity and molality are equivalent. So,

$$\Delta T_f = k_f m = (1.86 \text{ °Cm}^{-1})(4.98 \times 10^{-5} \text{ m})$$

$$= 9.26 \times 10^{-5} \text{ °C}$$

(d) The magnitude of the temperature change is too small to measure.

12.106 (a) $\Delta T_b = k_b m = (0.51 \text{ °Cm}^{-1})(1.00 \text{ m})$
       $= 0.51°C$
       $\Delta T_b = 100.51 \text{ °C}$

(b) $\Delta T_b = i k_b m = (4)(0.51 \text{ °Cm}^{-1})(1.00 \text{ m})$
     $= 2.04 \text{ °C}$
     $\Delta T_b = 102.04 \text{ °C}$

(c) $i = 0.183°C/0.51°C = 0.36$

Chapter Thirteen

Practice Exercises

P13.1   From the coefficients in the balanced equation we see that, for every two moles of SO_2 that is produced, 2 moles of H_2S are consumed, three moles of O_2 are consumed, and two moles of H_2O are produced.

$$\text{Rate of disappearance of } O_2 = \left(\frac{3 \, \text{mol} \, O_2}{2 \, \text{mol} \, SO_2}\right)\left(\frac{0.30 \, \text{mol}}{L \, s}\right) = 0.45 \, \text{mol} \, L^{-1} \, s^{-1}$$

$$\text{Rate of disappearance of } H_2S = \left(\frac{2 \, \text{mol} \, H_2S}{2 \, \text{mol} \, SO_2}\right)\left(\frac{0.30 \, \text{mol}}{L \, s}\right) = 0.30 \, \text{mol} \, L^{-1} \, s^{-1}$$

P13.2   The rate of the reaction after 250 seconds have elapsed is equal to the slope of the tangent to the curve at 250 seconds. First draw the tangent, and then estimate its slope as follows, where A is taken to represent one point on the tangent, and B is taken to represent another point on the tangent:

$$\text{rate} = \left(\frac{A \, (\text{mol/L}) - B \, (\text{mol/L})}{A \, (s) - B \, (s)}\right) = \frac{\text{change in concentration}}{\text{change in time}}$$

A value near $9.4 \times 10^{-5} \, \text{mol} \, L^{-1} \, s^{-1}$ is correct.

P13.3   (a)   First use the given data in the rate law:

Rate = $k[HI]^2$
$2.5 \times 10^{-4} \, \text{mol} \, L^{-1} \, s^{-1} = k[5.58 \times 10^{-2} \, \text{mol/L}]^2$
$k = 8.0 \times 10^{-2} \, L \, \text{mol}^{-1} \, s^{-1}$

(b) $L \, \text{mol}^{-1} \, s^{-1}$

P13.4   The order of the reaction with respect to a given substance is the exponent to which that substance is raised in the rate law:
order of the reaction with respect to $[BrO_3^-]$ = 1
order of the reaction with respect to $[SO_3^{2-}]$ = 1
overall order of the reaction = 1 + 1 = 2

P13.5   In each case, $k = \text{rate}/[A][B]^2$, and the units of k are $L^2 \, \text{mol}^{-2} \, s^{-1}$.

Each calculation is performed as follows, using the second data set as the example:

$$k = \frac{0.40 \, \text{mol} \, L^{-1} \, s^{-1}}{\left(0.20 \, \text{mol} \, L^{-1}\right)\left(0.10 \, \text{mol} \, L^{-1}\right)^2} = 2.0 \times 10^2 \, L^2 \, \text{mol}^{-2} \, s^{-1}$$

Each of the other data sets also gives the value:
$k = 2.0 \times 10^2 \, L^2 \, \text{mol}^{-2} \, s^{-1}$.

P13.6   The rate will likely take the form rate = $k[C_{12}H_{22}O_{11}]^n$, where n is the order of the reaction with respect to sucrose. A comparison of the first two lines of data shows that increasing the sucrose concentration by a factor of 2 (from 0.10 to 0.20) causes rate to increase also by nearly a factor of two (from 6.17×10^{-5} to 12.3×10^{-5}). This corresponds to the case in Table 13.3 for which a concentration increase by a factor of 2 causes a rate increase by a factor of $2 = 2^1$, and we conclude that the order of the reaction with respect to sucrose is one.

The same conclusion is reached on examining the data of the first and third rows. Increasing the concentration by a factor of 5 causes an increase in the rate by a factor of $5 = 5^1$, and we conclude that n = 1.

P13.7   The rate law will likely take the form rate = $k[A]^n[B]^{n'}$, where n and n' are the order of the reaction with respect to A and B, respectively. On comparing the first two lines of data, in which the concentration of B is held constant, we note that increasing the concentration of A by a factor of 2 (from 0.40 to 0.80) causes an increase in the rate by a factor of 4 (from 1.0×10^{-4} to 4.0×10^{-4}). Thus, we have a rate increase by 2^2, caused by a concentration increase by a factor of 2. This corresponds to the case in Table 13.3 for which n = 2, and we conclude that the reaction is second order with respect to A.

On comparing the second and third lines of data (wherein the concentration of A is held constant), we note that increasing the concentration of B by a factor of 2 (from 0.30 to 0.60) causes an increase in the rate by a factor of 4 (from 4.0×10^{-4} to 16.0×10^{-4}). This is an increase in rate by a factor of 2^2, and it forces us to conclude, using the information of Table 13.3, that the value of n' is 2. Thus the reaction is also second order with respect to B. The rate law is then written:

Rate = $k[A]^2[B]^2$

P13.8   (a)   We substitute into equation 13.5, first converting the time to seconds:
t = 2 hr × 3600 s/hr = 7200 s

$$\ln \frac{[A]_0}{[A]_t} = kt$$

$$\text{antiln} \left[\ln \frac{[A]_0}{[A]_t} \right] = \text{antiln} [kt] = \text{antiln} \left[\left(6.2 \times 10^{-5} \text{ s}^{-1} \right) \left(7200 \text{ s} \right) \right]$$

$$\frac{[A]_0}{[A]_t} = 1.56 = \frac{0.40 \text{ M}}{[A]_t}$$

$$[A]_t = \frac{0.40 \text{ M}}{1.56} = 0.26 \text{ M}$$

(b)   Again, we use equation 13.5, this time solving for time:

$$\ln \frac{[A]_0}{[A]_t} = kt$$

$$t = \frac{1}{k} \times \ln \frac{[A]_0}{[A]_t} = \frac{1}{6.2 \times 10^{-5} \text{ s}^{-1}} \times \ln \frac{0.40 \text{ M}}{0.30 \text{ M}} = 4600 \text{ s}$$

$4.6 \times 10^3 \text{ s} \times 1 \text{ min}/60 \text{ s} = 77 \text{ min}$

P13.9   This is a second order reaction, and we use equation 13.6:

$$\frac{1}{[NOCl]_t} - \frac{1}{[NOCl]_0} = kt$$

$$\frac{1}{[0.010 \text{ M}]} - \frac{1}{[0.040 \text{ M}]} = \left(0.020 \text{ L mol}^{-1} \text{ s}^{-1} \right) \times t$$

t = 3.8×10^3 s
t = 3.8×10^3 s × 1 min/60 s = 63 min

P13.10   For a first order reaction:

$$t_{1/2} = \frac{0.693}{k} = \frac{0.693}{6.17 \times 10^{-4} \, s^{-1}} = 1.12 \times 10^3 \, s$$

$$t_{1/2} = 1.12 \times 10^3 \, s \times \frac{1 \, min}{60 \, s} = 18.7 \, min$$

If we refer to the chart given in the text in example 13.9, we see that two half lives will have passed if there is to be only one quarter of the original amount of material remaining. This corresponds to:

18.7 min per half life × 2 half lives = 37.4 min

P13.11   The reaction is first order. A second order reaction should have a half life that depends on the initial concentration according to equation 13.8.

P13.12   (a)   Use equation 13.11.

$$\ln\frac{k_2}{k_1} = \frac{-E_a}{R}\left[\frac{1}{T_2} - \frac{1}{T_1}\right]$$

$$\ln\left[\frac{23 \, L \, mol^{-1} \, s^{-1}}{3.2 \, L \, mol^{-1} \, s^{-1}}\right] = \frac{-E_a}{8.314 \, J \, mol^{-1} \, K^{-1}}\left[\frac{1}{673 \, K} - \frac{1}{623 \, K}\right]$$

Solving for E_a gives 1.4×10^5 J/mol = 1.4×10^2 kJ/mol

(b)   We again use equation 13.11, substituting the values:
$k_1 = 3.2 \, L \, mol^{-1} \, s^{-1}$ \qquad at T_1 = 623 K
$k_2 = ?$ \qquad\qquad\qquad at T_2 = 573 K

$$\ln\frac{k_2}{k_1} = \frac{-E_a}{R}\left[\frac{1}{T_2} - \frac{1}{T_1}\right]$$

$$\ln\left[\frac{k_2}{3.2 \, L \, mol^{-1} \, s^{-1}}\right] = \frac{-1.4 \times 10^5 \, J \, mol^{-1}}{8.314 \, J \, mol^{-1} \, K^{-1}}\left[\frac{1}{573 \, K} - \frac{1}{623 \, K}\right]$$

Solving for k_2 gives 0.30 L mol^{-1} s^{-1}.

P13.13   If it is known that the entire process occurs by a single, elementary mechanistic step, then we have the special circumstance that the order of the reaction with respect to each chemical participating in the elementary step is given directly by the coefficient of that chemical in the balanced equation for the elementary step:

Rate = $k[NO]^1[O_3]^1$

P13.14   The slow step (second step) of the mechanism determines the rate law:
Rate = $k[NO_2Cl]^1[Cl]^1$
However, Cl is an intermediate and cannot be part of the rate law expression. We need to solve for the concentration of Cl by using the first step of the mechanism. Assuming that the first step is an equilibrium, the rates of the forward and reverse reactions are equal:
Rate = $k_{forward}[NO_2Cl] = k_{reverse}[Cl][NO_2]$
Solving for [Cl] we get
$$[Cl] = \frac{k_f}{k_r}\frac{[NO_2Cl]}{[NO_2]}$$
Substituting into the rate law expression for the second step yields:

$$Rate = \frac{k[NO_2Cl]^2}{[NO_2]}$$, where all the constants have been combined into one new constant.

Review Problems

13.56   The two rates of disappearance of SO_2Cl_2 are given by the slopes of tangents to the curve at the designated times:
Rate at t = 200 min = 1.0×10^{-4} moles L^{-1} min^{-1}
Rate at t = 600 min = 5.9×10^{-5} moles L^{-1} min^{-1}

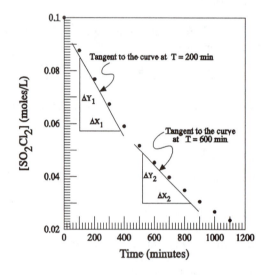

13.58   This is determined by the coefficients of the balanced chemical equation. For every mole of N_2 that reacts, 3 mol of H_2 will react. Thus the rate of disappearance of hydrogen is three times the rate of disappearance of nitrogen. Similarly, the rate of disappearance of N_2 is half the rate of appearance of NH_3, or NH_3 appears twice as fast as N_2 disappears.

13.60   (a)   rate for O_2 = -1.20 mol L^{-1} s^{-1} $\times 19/2$ = -11.4 mol L^{-1} s^{-1}
         (b)   rate for CO_2 = $+1.20$ mol L^{-1} s^{-1} $\times 12/2$ = 7.20 mol L^{-1} s^{-1}
         (c)   rate for H_2O = $+1.20$ mol L^{-1} s^{-1} $\times 14/2$ = 8.40 mol L^{-1} s^{-1}

13.62   rate = $(7.1 \times 10^9$ L^2 mol^{-2} $s^{-1})(1.0 \times 10^{-3}$ $mol/L)^2(3.4 \times 10^{-2}$ $mol/L)$
         rate = 2.4×10^2 mol L^{-1} s^{-1}

13.64   In each case, the order with respect to a reactant is the exponent to which that reactant's concentration is raised in the rate law.
         (a)   For $HCrO_4^-$, the order is 1.
                For HSO_3^-, the order is 2.
                For H^+, the order is 1.

         (b)   The overall order is $1 + 2 + 1 = 4$.

13.66   On comparing the data of the first and second experiments, we find that, whereas the concentration of N is unchanged, the concentration of M has been doubled, causing a doubling of the rate. This corresponds to the fourth case in Table 13.3, and we conclude that the order of the reaction with respect to M is 1. In the second and third experiments, we have a different result. When the concentration of M is held constant, the concentration of N is tripled, causing an increase in the rate by a factor of nine. This constitutes the eighth case in Table 13.3, and we conclude that the order of the reaction with respect to N is 2. This means that the overall rate expression is: rate = k[M][N]^2 and we can solve for the value of k by substituting the appropriate data:

5.0 X 10^{-3} mol L^{-1} s^{-1} = k × [0.020 mol/L][0.010 mol/L]^2
k = 2.5 X 10^3 L^2 mol^{-2} s^{-1}

13.68 The reaction is first order in OCl$^-$, because an increase in concentration by a factor of two, while holding the concentration of I$^-$ constant (compare the first and second experiments of the table), has caused an increase in rate by a factor of 2^1 = 2. The order of reaction with respect to I$^-$ is also 1, as is demonstrated by a comparison of the first and third experiments.

rate = k[OCl$^-$][I$^-$]
Using the last data set:
3.5 X 10^4 mol L^{-1} s^{-1} = k[1.7 X 10^{-3} mol/L][3.4 X 10^{-3} mol/L]
k = 6.1 X 10^9 L mol^{-1} s^{-1}

13.70 Compare the first and second experiments. On doubling the ICl concentration, the rate is found to increase by a factor of 2 = 2^1, and the order of the reaction with respect to ICl is 1 (case number four in Table 13.3). In the first and third experiments, the concentration of ICl is constant, whereas the concentration of H$_2$ in the first experiment is twice that in the third. This causes a change in the rate by a factor of 2 also, and the rate law is found to be: rate = k[ICl][H$_2$]. Using the data of the first experiment:
1.5 X 10^{-3} mol L^{-1} s^{-1} = k[0.10 mol L^{-1}][0.10 mol L^{-1}]
k = 1.5 X 10^{-1} L mol^{-1} s^{-1}

13.72 A graph of ln [SO$_2$Cl$_2$]$_t$ versus t will yield a straight line if the data obeys a first–order rate law.

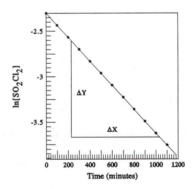

These data do yield a straight line when ln [SO$_2$Cl$_2$]$_t$ is plotted against the time, t. The slope of this line equals –k. Plotting the data provided and using linear regression to fit the data to a straight line yields a value of 1.32 X 10^{-3} min^{-1} for k.

13.74 (a) The time involved must be converted to a value in seconds:
1 hr × 3600 s/hr = 3.6 X 10^3 s, and then we make use of equation 13.5, where x is taken to represent the desired SO$_2$Cl$_2$ concentration:

$$\ln \frac{0.0040\,M}{x} = (2.2 \times 10^{-5}\,s^{-1})(3.6 \times 10^3\,s)$$

x = 3.7 X 10^{-3} M

(b) The time is converted to a value having the units seconds
24 hr × 3600 s/hr = 8.64 X 10^4 s, and then we use equation 13.5, where x is taken to represent the desired SO$_2$Cl$_2$ concentration:

$$\ln \frac{0.0040\,M}{x} = (2.2 \times 10^{-5}\,s^{-1})(8.64 \times 10^{4}\,s)$$

$$x = 6.0 \times 10^{-4}\,M$$

13.76 Any consistent set of units for expressing concentration may be used in equation 13.5, where we let A represent the drug that is involved:

$$\ln \frac{[A]_0}{[A]_t} = kt$$

$$\ln \frac{25.0\,^{mg}/_{kg}}{15.0\,^{mg}/_{kg}} = k(120\,\text{min})$$

Solving for k we get $4.26 \times 10^{-3}\,\text{min}^{-1}$

13.78 We use the equation:

$$\frac{1}{[HI]_t} - \frac{1}{[HI]_0} = kt$$

$$\frac{1}{[8.0 \times 10^{-4}\,M]} - \frac{1}{[3.4 \times 10^{-2}\,M]} = \left(1.6 \times 10^{-3}\,L\,mol^{-1}\,s^{-1}\right) \times t$$

Solving for t gives:
t = 7.6×10^{5} s or t = 7.6×10^{5} s $\times$ 1 min/60 s = 1.3×10^{4} min

13.80 # half lives = $(2.0\,\text{hrs})\left(\dfrac{60\,\text{min}}{1\,\text{hr}}\right)\left(\dfrac{1\,\text{half life}}{15\,\text{min}}\right) = 8.0$ half lives

Eight half lives correspond to the following fraction of original material remaining:

Number of half lives	Fraction remaining
1	1/2
2	1/4
3	1/8
4	1/16
5	1/32
6	1/64
7	1/128
8	1/256

13.82 It requires approximately 500 min (as determined from the graph) for the concentration of SO_2Cl_2 to decrease from 0.100 M to 0.050 M, i.e., to decrease to half its initial concentration. Likewise, in another 500 minutes, the concentration decreases by half again, i.e. from 0.050 M to 0.025 M. This means that the half life of the reaction is independent of the initial concentration, and we conclude that the reaction is first order in SO_2Cl_2.

13.84 t = 0.693/k = 0.693/1.6 $\times 10^{-3}\,s^{-1}$ = 4.3×10^{2} seconds

13.86 The graph is prepared exactly as in an example of the text. The slope is found using linear regression, to be: -9.5×10^{3} K. Thus -9.5×10^{3} K = $-E_a/R$
$E_a = -(-9.5 \times 10^{3}\,K)(8.314\,J\,K^{-1}\,mol^{-1}) = 7.9 \times 10^{4}$ J/mol = 79 kJ/mol

Using the equation, we proceed as follows:

$$\ln\frac{k_2}{k_1} = \frac{-E_a}{R}\left[\frac{1}{T_2} - \frac{1}{T_1}\right]$$

$$\ln\left[\frac{1.94 \times 10^{-3}\ \text{L mol}^{-1}\text{s}^{-1}}{2.88 \times 10^{-4}\ \text{L mol}^{-1}\text{s}^{-1}}\right] = \frac{-E_a}{8.314\ \text{J mol}^{-1}\text{K}^{-1}}\left[\frac{1}{673\ \text{K}} - \frac{1}{593\ \text{K}}\right]$$

$$1.907 = \frac{2.00 \times 10^{-4}\ \text{K}^{-1}}{8.314\ \text{J mol}^{-1}\text{K}^{-1}} \times E_a$$

$$E_a = 7.93 \times 10^4\ \text{J/mol} = 79.3\ \text{kJ/mol}$$

13.88   Using the equation we have:

$$\ln\frac{k_2}{k_1} = \frac{-E_a}{R}\left[\frac{1}{T_2} - \frac{1}{T_1}\right]$$

$$\ln\left[\frac{1.0 \times 10^{-3}\ \text{L mol}^{-1}\text{s}^{-1}}{9.3 \times 10^{-5}\ \text{L mol}^{-1}\text{s}^{-1}}\right] = \frac{-E_a}{8.314\ \text{J mol}^{-1}\text{K}^{-1}}\left[\frac{1}{403\ \text{K}} - \frac{1}{373\ \text{K}}\right]$$

$$2.37 = \frac{2.00 \times 10^{-4}\ \text{K}^{-1}}{8.314\ \text{J mol}^{-1}\text{K}^{-1}} \times E_a$$

$$E_a = 9.89 \times 10^4\ \text{J/mol} = 98.9\ \text{kJ/mol}$$

Equation states $k = A\exp\left(\dfrac{-E_a}{RT}\right)$

$$A = \frac{k}{\exp\left(\dfrac{-E_a}{RT}\right)}$$

$$= \frac{9.3 \times 10^{-5}\ \text{L mol}^{-1}\text{s}^{-1}}{\exp\left(\dfrac{-9.89 \times 10^4\ \text{J}/_{\text{mol}}}{\left(8.314\ \text{J}/_{\text{mol K}}\right)(373\ \text{K})}\right)}$$

$$= 6.6 \times 10^9\ \text{L mol}^{-1}\text{s}^{-1}$$

13.90   Substituting into the equation:

$$\ln\frac{k_2}{k_1} = \frac{-E_a}{R}\left[\frac{1}{T_2} - \frac{1}{T_1}\right]$$

$$\ln\left[\frac{3.75 \times 10^{-2}\ \text{s}^{-1}}{2.1 \times 10^{-3}\ \text{s}^{-1}}\right] = \frac{-E_a}{8.314\ \text{J mol}^{-1}\text{K}^{-1}}\left[\frac{1}{298\ \text{K}} - \frac{1}{273\ \text{K}}\right]$$

$$(3.70 \times 10^{-5}\ \text{mol/J})(E_a) = 2.88$$
$$E_a = 7.8 \times 10^4\ \text{J/mol} = 78\ \text{kJ/mol}$$

13.92   We can use the equation:

(a)

$$k = A \exp\left(\frac{-E_a}{RT}\right)$$

$$= \left(4.3 \times 10^{13} \text{ s}^{-1}\right) \exp\left(\frac{-103 \times 10^3 \text{ J mol}^{-1}}{\left(8.314 \text{ }^{J}/_{\text{mol K}}\right)\left(293 \text{ K}\right)}\right)$$

$$= 1.9 \times 10^{-5} \text{ s}^{-1}$$

(b)

$$k = A \exp\left(\frac{-E_a}{RT}\right)$$

$$= \left(4.3 \times 10^{13} \text{ s}^{-1}\right) \exp\left(\frac{-103 \times 10^3 \text{ J mol}^{-1}}{\left(8.314 \text{ }^{J}/_{\text{mol K}}\right)\left(373 \text{ K}\right)}\right)$$

$$= 1.6 \times 10^{-1} \text{ s}^{-1}$$

13.95 $k = 0.693/12.5 \text{ y} = 5.54 \times 10^{-2} \text{ y}^{-1}$

$$\ln\frac{1}{0.1} = \left(5.54 \times 10^{-2} \text{ y}^{-1}\right) \times t$$

$t = 41.6 \text{ y}$

13.97   (a) $\text{rate} = k_1[A]^2$
        (b) $\text{rate} = k_{-1}[A_2]^1$
        (c) $\text{rate} = k_2[A_2]^1[E]^1$
        (d) $2A + E \rightarrow B + C$
        (e)   The rates for the forward and reverse directions of step one are set equal to each other in order to arrive at an expression for the intermediate $[A_2]$ in terms of the reactant $[A]$: $k_1[A]^2 = k_{-1}[A_2]$

$$[A_2] = \frac{k_1}{k_{-1}}[A]^2$$

This is substituted into the rate law for question (c) above, giving a rate expression that is written using only observable reactants: $\text{rate} = k_2 \dfrac{k_1}{k_{-1}}[A]^2[E]^1$

13.99   First, determine a value for E_a using the equation:

$$\ln\frac{k_2}{k_1} = \frac{-E_a}{R}\left[\frac{1}{T_2} - \frac{1}{T_1}\right]$$

$$\ln\left[\frac{2.25 \times 10^{-5} \text{ min}^{-1}}{5.84 \times 10^{-6} \text{ min}^{-1}}\right] = \frac{-E_a}{8.314 \text{ J mol}^{-1} \text{ K}^{-1}}\left[\frac{1}{343 \text{ K}} - \frac{1}{333 \text{ K}}\right]$$

$$1.35 = \frac{8.76 \times 10^{-5} \text{ K}^{-1}}{8.314 \text{ J mol}^{-1} \text{ K}^{-1}} \times E_a$$

$E_a = 1.28 \times 10^5 \text{ J/mol} = 128 \text{ kJ/mol}$

Next, use this value of E_a and the data at 70 °C to calculate a rate constant at 80 °C:

$$\ln \frac{k_2}{k_1} = \frac{-E_a}{R}\left[\frac{1}{T_2} - \frac{1}{T_1}\right]$$

$$\ln\left[\frac{k_2}{2.25 \times 10^{-5}\,\text{min}^{-1}}\right] = \frac{-1.28 \times 10^5\,\text{J mol}^{-1}}{8.314\,\text{J mol}^{-1}\,\text{K}^{-1}}\left[\frac{1}{353\,\text{K}} - \frac{1}{343\,\text{K}}\right]$$

$$\ln\left[\frac{k_2}{2.25 \times 10^{-5}\,\text{min}^{-1}}\right] = 1.27$$

$k_2 = 2.25 \times 10^{-5}\,\text{min}^{-1} \times \exp(1.27) = 8.02 \times 10^{-5}\,\text{min}^{-1}$.

Finally, use the first order rate expression to determine time:

$$\ln\frac{0.0020\,\text{M}}{0.0012\,\text{M}} = 8.02 \times 10^{-5}\,\text{min}^{-1} \times t$$

Solving for t we get 6.34×10^3 min.

13.102

$$\ln \frac{k_2}{k_1} = \frac{-E_a}{R}\left[\frac{1}{T_2} - \frac{1}{T_1}\right]$$

$$\ln\left[\frac{2}{1}\right] = \frac{-E_a}{8.314\,\text{J mol}^{-1}\,\text{K}^{-1}}\left[\frac{1}{308\,\text{K}} - \frac{1}{298\,\text{K}}\right]$$

$(1.31 \times 10^{-5}\,\text{mol/J})(E_a) = 0.693$

$E_a = 5.29 \times 10^4\,\text{J/mol} = 52.9\,\text{kJ/mol}$

13.104 (a) The number of chirps in eight seconds is simply the temperature minus 4. So, in order, there will be 16, 21, 26 and 31 chirps at these temperatures.

(b) The data are plotted below.

# of chirps	T	1/T	ln (chirps)
16	293	0.003412	2.772588
21	298	0.003355	3.044522
26	303	0.003300	3.258096
31	308	0.003246	3.433987

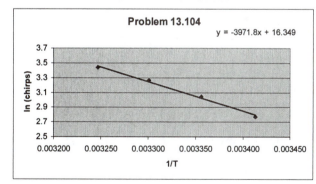

The slope of the line is equal to the Activation energy divided by R so $E_a = 477.4\,\text{J/mol}$.

(c) At 120 °C the cricket is most likely dead and, hence, there would be no chirps.

13.107 (a) The first step, in which a free radical is produced, is the initiation step.

(b)   Both the second and third steps are propagating steps since HBr, the desired product, and an additional free radical are produced.

(c)   The final step in which two bromine free radicals recombine to give a bromine molecule is the termination step. The presence of the additional reaction step serves to decrease the concentration of HBr.

Practice Exercises

P14.1 (a) $K_c = \dfrac{[H_2O]^2}{[H_2]^2[O_2]}$

(b) $K_c = \dfrac{[NH_4^+][OH^-]}{[NH_3][H_2O]}$

P14.2 Since the starting equation has been reversed and divided by two, we must invert the equilibrium constant, and then take the square root: $K_c = 1.2 \times 10^{-13}$

P14.3 If we divide both equations by 2 and reverse the second we get:

$CO(g) + 1/2 O_2(g) \rightarrow CO_2(g) \qquad K_c = 5.7 \times 10^{45}$
$H_2O(g) \rightarrow H_2(g) + 1/2 O_2(g) \qquad K_c = 3.3 \times 10^{-41}$

Note that when we divide the equation by two, we need to take the square root of the rate constant. When we reverse the reaction, we need to take the inverse.

Adding these equations we get the desired equation so we need simply multiply the values for K_c in order to obtain the new value: $K_c = 1.9 \times 10^5$

P14.4 $K_c = \dfrac{(P_{HI})^2}{(P_{H_2})(P_{I_2})}$

P14.5 K_c is a measure of the ratio of the amount of products to the amount of reactants at equilibrium. Consequently, a large value of K_c indicates a large amount of products. The reaction that will proceed furthest to completion will have the largest value of K_c. For these reactions, (b) will proceed farthest to completion.

P14.6 Use the equation:

$$K_p = K_c(RT)^{\Delta n_g}$$
$$K_c = \dfrac{K_p}{(RT)^{\Delta n_g}}$$

In this case, $\Delta n_g = (1 - 3) = -2$, and we have:

$$K_c = \dfrac{K_p}{(RT)^{\Delta n_g}} = \dfrac{3.8 \times 10^{-2}}{\left((0.0821 \frac{L\,atm}{mol\,K})(473\,K)\right)^{-2}} = 57$$

P14.7 Use the equation

$$K_p = K_c(RT)^{\Delta n_g}$$

In this reaction, $\Delta n_g = 3 - 2 = 1$

$$K_p = K_c(RT)^{\Delta n_g} = \left(7.3 \times 10^{34}\right)\left((0.0821 \tfrac{L\,atm}{mol\,K})(298\,K)\right)^1 = 1.8 \times 10^{36}$$

P14.8 (a) $\quad K_c = \dfrac{1}{[Cl_2 \, (g)]}$

(b) $\quad K_c = \dfrac{1}{[NH_3 \, (g)][HCl(g)]}$

(c) $\quad K_c = [Ag^+]^2[CrO_4^{2-}]$

P14.9 (a) The equilibrium will shift to the right, decreasing the concentration of Cl_2 at equilibrium, and consuming some of the added PCl_3. The value of K_p will be unchanged.

(b) The equilibrium will shift to the left, consuming some of the added PCl_5 and increasing the amount of Cl_2 at equilibrium. The value of K_p will be unchanged.

(c) For any exothermic equilibrium, an increase in temperature causes the equilibrium to shift to the left, in order to remove energy in response to the stress. This equilibrium is shifted to the left, making more Cl_2 and more PCl_3 at the new equilibrium. The value of K_p is given by the following:

$$K_p = \frac{P_{PCl_5}}{P_{PCl_3} \times P_{Cl_2}}$$

In this system, an increase in temperature (which causes an increase in the equilibrium concentrations of both PCl_3 and Cl_2 and a decrease in the equilibrium concentration of PCl_5) causes an increase in the denominator of the above expression as well as a decrease in the numerator of the above expression. Both of these changes serve to decrease the value of K_p.

(d) Decreasing the container volume for a gaseous system will produce an increase in partial pressures for all gaseous reactants and products. In order to lower the increase in partial pressures, the equilibrium will shift so as to favor the reaction side having the smaller number of gaseous molecules, in this case to the right. This shift will decrease the amount of Cl_2 and PCl_3 at equilibrium, and it will increase the amount of PCl_5 at equilibrium. This increases the size of the numerator and decreases the size of the denominator in the above expression for K_p, causing the value of K_p to increase.

P14.10 $\quad K_c = \dfrac{[CO_2][H_2]}{[CO][H_2O]} = \dfrac{(0.150)(0.200)}{(0.180)(0.0411)} = 4.06$

P14.11 $\quad 2CO(g) + O_2(g) \rightarrow 2CO_2(g)$

Using the stoichiometry of the reaction we can see that for every mol of O_2 that is used, twice as much CO will react and twice as much CO_2 will be produced. Consequently, if the $[O_2]$ decreases by 0.030 mol/L, the $[CO]$ decreases by 0.060 mol/L and $[CO_2]$ increases by 0.060 mol/L.

P14.12 (a) The initial concentrations were:
$[PCl_3]$ = 0.200 mol/1.00 L = 0.20 M
$[Cl_2]$ = 0.100 mol/1.00 L = 0.100 M
$[PCl_5]$ = 0.00 mol/1.00 L = 0.000 M

(b) The change in concentration of PCl_3 was (0.200 − 0.120) M = 0.080 mol/L. The other materials must have undergone changes in concentration that are dictated by the coefficients of the balanced chemical equation, namely: $PCl_3 + Cl_2 \rightarrow PCl_5$ or both PCl_3 and Cl_2 have decreased by 0.080 M and PCl_5 has increased by 0.080 M.

(c) As stated in the problem, the equilibrium concentration of PCl_3 is 0.120 M. The equilibrium concentration of PCl_5 is 0.080 M since initially there was no PCl_5. The equilibrium concentration of Cl_2 equals the initial concentration minus the amount that reacted, 0.100 M - 0.080 M = 0.020 M.

(d) $\quad K_c = \dfrac{[PCl_5]}{[PCl_3][Cl_2]} = \dfrac{(0.080)}{(0.120)(0.020)} = 33$

P14.13 $K_c = \dfrac{[CH_3CO_2C_2H_5][H_2O]}{[CH_3CO_2H][C_2H_5OH]} = \dfrac{(0.910)(0.00850)}{(0.210)[C_2H_5OH]} = 4.10$

$[C_2H_5OH] = 8.98 \times 10^{-3}$ M

P14.14 Initially we have $[H_2] = [I_2] = 0.200$ M.

	$[H_2]$	$[I_2]$	$[HI]$
I	0.200	0.200	–
C	–x	–x	+2x
E	0.200–x	0.200–x	+2x

Substituting the above values for equilibrium concentrations into the mass action expression gives:

$$K_c = \frac{[HI]^2}{[H_2][I_2]} = \frac{(2x)^2}{(0.200 - x)(0.200 - x)} = 49.5$$

Take the square root of both sides of this equation to get; $\dfrac{2x}{(0.200 - x)} = 7.04$. This equation is easily solved giving x = 0.156. The substances then have the following concentrations at equilibrium: $[H_2] = [I_2] = 0.200 - 0.156 = 0.044$ M, $[HI] = 2(0.156) = 0.312$ M.

P14.15 $N_2(g) + O_2(g) \rightleftharpoons 2NO(g)$

	$[N_2]$	$[O_2]$	$[NO]$
I	0.033	0.00810	–
C	–x	–x	+2x
E	0.033–x	0.00810–x	+2x

Substituting the above values for equilibrium concentrations into the mass action expression gives:

$$K_c = \frac{[NO]^2}{[N_2][O_2]} = \frac{(2x)^2}{(0.033 - x)(0.00810 - x)} = 4.8 \times 10^{-31}$$

If we assume that x << 0.033 and x << 0.00810, we can simplify this equation. (Because the value of K_c is so low, this assumption should be valid.) The equation simplifies as:

$$K_c = \frac{(2x)^2}{(0.033)(0.00810)} = 4.8 \times 10^{-31}$$

This equation is easily solved to give $x = 5.7 \times 10^{-18}$ M. The equilibrium concentration of NO is 2x according to the ICE table so, $[NO] = 1.1 \times 10^{-17}$ M.

Review Problems

14.19 (a) $K_c = \dfrac{[POCl_3]^2}{[PCl_3]^2[O_2]}$

(d) $K_c = \dfrac{[NO_2]^2[H_2O]^8}{[N_2H_4][H_2O_2]^6}$

(b) $K_c = \dfrac{[SO_2]^2[O_2]}{[SO_3]^2}$

(e) $K_c = \dfrac{[SO_2][HCl]^2}{[SOCl_2][H_2O]}$

(c) $\quad K_c = \dfrac{[NO]^2 [H_2O]^2}{[N_2H_4][O_2]^2}$

14.21 (a) $\quad K_p = \dfrac{\left(P_{POCl_3}\right)^2}{\left(P_{PCl_3}\right)^2 \left(P_{O_2}\right)}$

(d) $\quad K_p = \dfrac{\left(P_{NO_2}\right)^2 \left(P_{H_2O}\right)^8}{\left(P_{N_2H_4}\right)\left(P_{H_2O_2}\right)^6}$

(b) $\quad K_p = \dfrac{\left(P_{SO_2}\right)^2 \left(P_{O_2}\right)}{\left(P_{SO_3}\right)^2}$

(e) $\quad K_p = \dfrac{\left(P_{SO_2}\right)\left(P_{HCl}\right)^2}{\left(P_{SOCl_2}\right)\left(P_{H_2O}\right)}$

(c) $\quad K_p = \dfrac{\left(P_{NO}\right)^2 \left(P_{H_2O}\right)^2}{\left(P_{N_2H_4}\right)\left(P_{O_2}\right)^2}$

14.23 (a) $\quad K_c = \dfrac{\left[Ag(NH_3)_2^+\right]}{\left[Ag^+\right]\left[NH_3\right]^2}$

(b) $\quad K_c = \dfrac{\left[Cd(SCN)_4^{2-}\right]}{\left[Cd^{2+}\right]\left[SCN^-\right]^4}$

14.25 The first equation has been reversed in making the second equation. We therefore take the inverse of the value of the first equilibrium constant in order to determine a value for the second equilibrium constant: $K = 1 \times 10^{85}$

14.27 (a) $\quad K_c = \dfrac{[HCl]^2}{[H_2][Cl_2]}$ (b) $\quad K_c = \dfrac{[HCl]}{[H_2]^{1/2}[Cl_2]^{1/2}}$

K_c for reaction (b) is the square root of K_c for reaction (a).

14.29 $M = P/RT$

$$M = \dfrac{(745\,\text{torr})\left(\dfrac{1\,\text{atm}}{760\,\text{torr}}\right)}{\left(0.0821\,\frac{L\,atm}{mol\,K}\right)(318\,K)} = 0.0375\,M$$

14.31 b

14.33 $K_p = K_c \times (RT)^{\Delta n_g}$
$6.3 \times 10^{-3} = K_c[(0.0821\ L\ atm\ K^{-1}\ mol^{-1})(498\ K)]^{-2} = 5.98 \times 10^{-4} \times K_c$
$K_c = 11$

14.35 $K_p = K_c \times (RT)^{\Delta n_g}$
$K_p = 4.2 \times 10^{-4}[(0.0821\ L\ atm\ K^{-1}\ mol^{-1})(773\ K)]^1 = 2.7 \times 10^{-2}$

14.37 $K_p = K_c \times (RT)^{\Delta n_g}$
$K_p = (0.40)[(0.0821\ L\ atm\ K^{-1}\ mol^{-1})(1046\ K)]^{-2} = 5.4 \times 10^{-5}$

14.39   In each case we get approximately 55.5 M:
        (a)

$$\# \, mol \, H_2O = \left(18.0 \, mL \, H_2O\right)\left(\frac{1\,g}{1\,mL}\right)\left(\frac{1\,mol\,H_2O}{18.02\,g\,H_2O}\right) = 0.999 \, mol \, H_2O$$

$$M = \left(\frac{0.999\,mol\,H_2O}{18.0\,mL\,H_2O}\right)\left(\frac{1000\,mL}{1\,L}\right) = 55.5 \, M$$

        (b)

$$\# \, mol \, H_2O = \left(100.0 \, mL \, H_2O\right)\left(\frac{1\,g}{1\,mL}\right)\left(\frac{1\,mol\,H_2O}{18.02\,g\,H_2O}\right) = 5.549 \, mol \, H_2O$$

$$M = \left(\frac{5.549\,mol\,H_2O}{100.0\,mL\,H_2O}\right)\left(\frac{1000\,mL}{1\,L}\right) = 55.49 \, M$$

        (c)

$$\# \, mol \, H_2O = \left(1.00 \, L \, H_2O\right)\left(\frac{1000\,mL}{1\,L}\right)\left(\frac{1\,g}{1\,mL}\right)\left(\frac{1\,mol\,H_2O}{18.02\,g\,H_2O}\right)$$

$$= 55.5 \, mol \, H_2O$$

$$M = \left(\frac{55.5\,mol\,H_2O}{1.00\,L\,H_2O}\right) = 55.5 \, M$$

14.41   (a) $K_c = \dfrac{[CO]^2}{[O_2]}$ (c) $K_c = \dfrac{[CH_4][CO_2]}{[H_2O]^2}$

        (b) $K_c = [H_2O][SO_2]$ (d) $K_c = \dfrac{[H_2O][CO_2]}{[HF]^2}$

        (e) $K_c = [H_2O]^5$

14.43

	[HCl]	[HI]	[Cl₂]
I	0.100	–	–
C	–2x	+2x	+x
E	0.100–2x	+2x	+x

Note: Since the $I_2(s)$ has a constant concentration, it may be neglected.

$$K_c = \frac{[HI]^2[Cl_2]}{[HCl]^2} = 1.0 \times 10^{-34}$$

$$K_c = \frac{(2x)^2(x)}{(0.100-2x)^2} = 1.0 \times 10^{-34}$$

Because the value of K_c is so small, we make the simplifying assumption that $(0.100 - 2x) \approx 0.100$, and the above equation becomes:

$$K_c = \frac{[HI]^2[Cl_2]}{[HCl]^2} = 1.0 \times 10^{-34}$$

$$K_c = \frac{(2x)^2(x)}{(0.100)^2} = 1.0 \times 10^{-34}$$

$4x^3 = 1.6 \times 10^{-36}$; $\therefore$ $x = 7.37 \times 10^{-13}$, and the above assumption is seen to have been valid.

$[HI] = 2x = 1.47 \times 10^{-12}$ M
$[Cl_2] = x = 7.37 \times 10^{-13}$ M
$[HCl] = (0.100 - 2x) \approx 0.100$ M

14.45   The mass action expression for this equilibrium is:

$$K_c = \frac{[PCl_5]}{[PCl_3][Cl_2]} = 0.18$$

and the value for the ion product for this system is:

$$Q = \frac{(0.00500)}{(0.0420)(0.0240)} = 4.96$$

(a)   This is not the value of the equilibrium constant, and we conclude that the system is not at equilibrium.

(b)   Since the value of the ion product for this system is larger than that of the equilibrium constant, the system must shift to the left to reach equilibrium.

14.47 $K_c = \dfrac{[CH_3OH]}{[CO][H_2]^2} = \dfrac{[CH_3OH]}{(0.180)(0.220)^2} = 0.500$

$[CH_3OH] = 4.36 \times 10^{-3}$ M.

14.49 $K_c = \dfrac{[CH_3OH]}{[CO][H_2]^2} = \dfrac{(0.00261)}{(0.105)(0.250)^2} = 0.398$

14.51

	[HBr]	[H_2]	[Br_2]
I	0.500	–	–
C	–2x	+x	+x
E	0.500–2x	+x	+x

The problem tell us that $[Br_2] = 0.0955$ M = x at equilibrium. Using the ICE table as a guide we see that the equilibrium concentrations are; $[H_2] = [Br_2] = 0.0955$ M and $[HBr] = 0.500 - 2(0.0955) = 0.309$ M.

$$K_c = \frac{[H_2][Br_2]}{[HBr]^2} = \frac{(0.0955)(0.0955)}{(0.309)^2} = 0.0955$$

14.53 According to the problem, the concentration of NO_2 increases in the course of this reaction. This means our ICE table will look like the following:

	$[NO_2]$	$[NO]$	$[N_2O]$	$[O_2]$
I	0.0560	0.294	0.184	0.377
C	+x	+x	−x	−x
E	0.0560+x	0.294+x	0.184−x	0.377−x

The problem tell us that $[NO_2]$ = 0.118 M = 0.0560+x at equilibrium. Solving we get; x = 0.062 M. Using the ICE table as a guide we see that the equilibrium concentrations are; [NO] = 0.356 M, $[N_2O]$ = 0.122 M and $[O_2]$ = 0.315 M.

$$K_c = \frac{[N_2O][O_2]}{[NO_2][NO]} = \frac{(0.122)(0.315)}{(0.118)(0.356)} = 0.915$$

14.55 $2BrCl \rightleftharpoons Br_2 + Cl_2$

	$[BrCl]$	$[Br_2]$	$[Cl_2]$
I	0.050	−	−
C	−2x	+x	+x
E	0.050−2x	+x	+x

Substituting the above values for equilibrium concentrations into the mass action expression gives:

$$K_c = \frac{[Br_2][Cl_2]}{[BrCl]^2} = \frac{(x)(x)}{(0.050 - 2x)^2} = 0.145$$

Take the square root of both sides to get

$$K_c = \frac{x}{0.050 - 2x} = 0.381$$

Solving for x gives: x = 0.011 M = $[Br_2]$ = $[Cl_2]$

14.57 The initial concentrations are each 0.240 mol/2.00 L = 0.120 M.

	$[SO_3]$	$[NO]$	$[NO_2]$	$[SO_2]$
I	0.120	0.120	−	−
C	−x	−x	+x	+x
E	0.120−x	0.120−x	+x	+x

Substituting the above values for equilibrium concentrations into the mass action expression gives:

$$K_c = \frac{[NO_2][SO_2]}{[SO_3][NO]} = \frac{(x)(x)}{(0.120 - x)(0.120 - x)} = 0.500$$

Taking the square root of both sides of this equation gives: 0.707 = x/(0.120 − x)
Solving for x we have: 1.707(x) = 0.0848, x = 0.0497 mol/L = $[NO_2]$ = $[SO_2]$, [NO] = $[SO_3]$ = 0.120 − x = 0.0703 mol/L

14.59 The initial concentrations are all 1.00 mol/100 L = 0.0100 M. Since the initial concentrations are all the same, the ion product is equal to 1.0, and we conclude that the system must shift to the left to reach equilibrium since $Q > K_c$.

	[CO]	[H₂O]	[CO₂]	[H₂]
I	0.0100	0.0100	0.0100	0.0100
C	+x	+x	−x	−x
E	0.0100+x	0.0100+x	0.0100−x	0.0100−x

Substituting the above values for equilibrium concentrations into the mass action expression gives:

$$K_c = \frac{[CO_2][H_2]}{[CO][H_2O]} = \frac{(0.0100-x)(0.0100-x)}{(0.0100+x)(0.0100+x)} = 0.400$$

We take the square root of both sides of the above equation:

$$\frac{(0.0100-x)}{(0.0100+x)} = 0.632$$

and $(0.632)(0.0100 + x) = 0.0100 - x$
$(1.632)x = 3.68 \times 10^{-3}$, or $x = 2.25 \times 10^{-3}$ mol/L The equilibrium concentrations are then: $[H_2] = [CO_2] = (0.0100 - 2.25 \times 10^{-3}) = 7.7 \times 10^{-3}$ M, $[CO] = [H_2O] = (0.0100 + 2.25 \times 10^{-3}) = 0.0123$ M.

14.61

	[HCl]	[H₂]	[Cl₂]
I	0.0500	–	–
C	−2x	+x	+x
E	0.0500−2x	+x	+x

Substituting the above values for equilibrium concentrations into the mass action expression gives:

$$K_c = \frac{[H_2][Cl_2]}{[HCl]^2} = \frac{(x)(x)}{(0.0500 - 2x)^2} = 3.2 \times 10^{-34}$$

Because K_c is so exceedingly small, we can make the simplifying assumption that x is also small enough to make $(0.0500 - 2x) \approx 0.0500$. Thus we have: $3.2 \times 10^{-34} = (x)^2/(0.0500)^2$

Taking the square root of both sides, and solving for the value of x gives:
$x = 8.9 \times 10^{-19}$ M $= [H_2] = [Cl_2]$
$[HCl] = (0.0500 - x) \approx 0.0500$ mol/L

14.63 $K_c = \frac{[CO]^2[O_2]}{[CO_2]^2} = 6.4 \times 10^{-7}$

	[CO₂]	[CO]	[O₂]
I	1.0 x 10⁻²	–	–
C	−2x	+2x	+x
E	1.0 x 10⁻² − 2x	+2x	+x

$K_c = \dfrac{[2x]^2[x]}{\left[1.0 \times 10^{-2} - 2x\right]^2} = 6.4 \times 10^{-7}$

Assume $x \ll 1.0 \times 10^{-2}$.

$\dfrac{4x^3}{(1.0 \times 10^{-2})^2} = 6.4 \times 10^{-7}$ $x = 2.5 \times 10^{-4}$

$[CO] = 2x = 5.0 \times 10^{-4}$ M

14.65 We first approach the problem in the normal fashion with an initial concentration of PCl_5 = 0.013 M.

	$[PCl_3]$	$[Cl_2]$	$[PCl_5]$
I	–	–	0.013
C	+x	+x	–x
E	+x	+x	0.013–x

Substituting the above values for equilibrium concentrations into the mass action expression gives:

$$K_c = \frac{[PCl_5]}{[PCl_3][Cl_2]} = \frac{(0.013-x)}{(x)(x)} = 0.18$$

rearranging; $0.18x^2 + x - 0.013 = 0$

We next attempt to use the quadratic equation to solve for the value of x, setting a = 0.18; b = 1; c = –0.013.

However, we find that unless we carry one more significant figure than is allowed, the quadratic formula for this problem gives us a concentration of zero for PCl_5. A better solution is obtained by "allowing" the initial equilibrium to shift **completely** to the left, giving us a new initial situation from which to work:

	$[PCl_3]$	$[Cl_2]$	$[PCl_5]$
I	0.013	0.013	–
C	–x	–x	+x
E	0.013–x	0.013–x	+x

Substituting the above values for equilibrium concentrations into the mass action expression gives:

$$K_c = \frac{[PCl_5]}{[PCl_3][Cl_2]} = \frac{(+x)}{(0.013-x)(0.013-x)} = 0.18$$

Now, we may assume that x << 0.013. The equation is simplified and we solve for x
$x = [PCl_5] = 3.0 \times 10^{-5}$ M.

14.67

	$[SO_3]$	$[NO]$	$[NO_2]$	$[SO_2]$
I	0.0500	0.100	–	–
C	–x	–x	+x	+x
E	0.0500–x	0.100–x	+x	+x

Substituting the above values for equilibrium concentrations into the mass action expression gives:

$$K_c = \frac{[NO_2][SO_2]}{[SO_3][NO]} = \frac{(x)(x)}{(0.0500-x)(0.100-x)} = 0.500$$

Since the equilibrium constant is not much larger than either of the values 0.0500 or 0.100, we cannot neglect the size of x in the above expression. A simplifying assumption is not therefore possible, and we must solve for the value of x using the quadratic equation. Multiplying out the above denominator, collecting like terms, and putting the result into the standard quadratic form gives:
$$0.500x^2 + (7.50 \times 10^{-2})x - (2.50 \times 10^{-3}) = 0$$

$$x = \frac{-7.50 \times 10^{-2} \pm \sqrt{\left(-7.50 \times 10^{-2}\right)^2 - 4(0.500)\left(-2.50 \times 10^{-3}\right)}}{2(0.500)} = 0.0281\,M,$$

using the (+) root. So, $[NO_2] = [SO_2] = 0.0281$ M

14.69 $\quad Kc = \dfrac{[CO][H_2O]}{[HCHO_2]^2} = 4.3 \times 10^5$

Since Kc is large, start by assuming all of the HCHO$_2$ decomposes to give CO and H$_2$O

	[HCHO$_2$]	[CO]	[O$_2$]
I	–	0.200	0.200
C	+x	–x	–x
E	+x	0.200 –x	0.200 –x

$$Kc = \frac{[0.200][0.200]}{[x]} = 4.3 \times 10^5$$

$x = 9.3 \times 10^{-8}$
so, at equilibrium
$[CO] = [H_2O] = 0.200 - x = 0.200$

Additional Exercises

14.71 (a) The mass action expression is:

$$K_p = \frac{\left(P_{NO_2}\right)^2}{\left(P_{N_2O_4}\right)} = 0.140\,atm$$

Solving the above expression for the partial pressure of NO$_2$, we get:

$$P_{NO_2} = \sqrt{P_{N_2O_4} \times K_p} = \sqrt{(0.250\,atm)(0.140\,atm)} = 0.187\,atm$$

(b) $\quad P_{total} = P_{NO_2} + P_{N_2O_4} = 0.187 + 0.250 = 0.437$ atm

14.73 The initial concentrations are:
$[NO_2] = (0.200\,mol/4.00\,L) = 0.0500$ M
$[NO] = (0.300\,mol/4.00\,L) = 0.0750$ M
$[N_2O] = (0.150\,mol/4.00\,L) = 0.0375$ M
$[O_2] = (0.250\,mol/4.00\,L) = 0.0625$ M

Substituting these values into the mass action expression we determine:
$$Q = \frac{[N_2O][O_2]}{[NO_2][NO]} = \frac{(0.0375)(0.0625)}{(0.0500)(0.0750)} = 0.625$$
Since Q<K$_c$, this reaction will proceed from left to right as written. The ICE table becomes:

	[NO$_2$]	[NO]	[N$_2$O]	[O$_2$]
I	0.0500	0.0750	0.0375	0.0625
C	–x	–x	+x	+x
E	0.0500–x	0.0750–x	0.0375+x	0.0625+x

Substituting the above values for equilibrium concentrations into the mass action expression gives:

$$K_c = \frac{[N_2O][O_2]}{[NO_2][NO]} = \frac{(0.0375 + x)(0.0625 + x)}{(0.0500 - x)(0.0750 - x)} = 0.914$$

To solve we need to use the quadratic equation. Expanding the above calculation we get:

$$0.086x^2 + 0.214x - 0.00109 = 0$$

Solving we get x = 0.00508. So,

$$[NO_2] = 0.0500 - x = 0.0449 M$$
$$[NO] = 0.0750 - x = 0.0699 \ M$$
$$[N_2O] = 0.0375 + x = 0.0426 \ M$$
$$[O_2] = 0.0625 + x = 0.0676 \ M$$

14.75   First, calculate a value for K_c using a rearranged form of equation 14.6, and setting the value for Δn to -1:

$$K_c = \frac{K_p}{(RT)^{\Delta n_g}} = \frac{1.5 \times 10^{18}}{\left[\left(0.0821 \frac{L \, atm}{mol \, K}\right)(300 \, K)\right]^{-1}} = 3.7 \times 10^{19}$$

The value for the equilibrium constant is very large, indicating that the equilibrium lies far to the right. We therefore anticipate that the initial conditions are unrealistic. The system is "allowed" to come to a more realistic **new initial** set of concentrations, by reaction of all of the starting amount of NO, to give a stoichiometric amount of N_2O and NO_2. Only then can we solve for the equilibrium concentrations in the usual manner:

	[NO]	[N₂O]	[NO₂]
I	–	0.010	0.010
C	+3x	–x	–x
E	+3x	0.010–x	0.010–x

Substituting the above values for equilibrium concentrations into the mass action expression gives:

$$Kc = \frac{[N_2O][NO_2]}{[NO]^3} = \frac{(0.010 - x)(0.010 - x)}{(3x)^3} = 3.7 \times 10^{19}$$

The simplifying assumption can be made that the value for x is much smaller than the number 0.010. Upon solving for x we get: $27x^3 = 2.7 \times 10^{-24}$ $x = 4.6 \times 10^{-9} \ M$
$[NO] = 3x = 1.4 \times 10^{-8} \ M$, $[N_2O] = [NO_2] = 0.010 - x = 0.010 \ M$.

14.77   First, calculate the number of moles of CO used in the experiment, using the ideal gas law:

$$n = \frac{PV}{RT} = \frac{(0.177 \, atm)(2.00 \, L)}{\left(0.0821 \frac{L \, atm}{mol \, K}\right)(298 \, K)} = 0.0145 \, mol$$

Next, calculate the partial pressure of CO at 400 C:

$$P = \frac{nRT}{V} = \frac{(0.0145 \, mol)\left(0.0821 \frac{L \, atm}{mol \, K}\right)(673 \, K)}{(2.00 \, L)} = 0.401 \, atm$$

Next, we calculate the number of moles of water that are supplied to the reaction, and convert to partial pressure for water, using the ideal gas equation:

$$\# \text{ mol H}_2\text{O} = (0.391 \text{ g H}_2\text{O})\left(\frac{1 \text{ mol H}_2\text{O}}{18.12 \text{ g H}_2\text{O}}\right) = 0.0217 \text{ mol H}_2\text{O}$$

$$P = \frac{nRT}{V} = \frac{(0.0217 \text{ mol})(0.0821 \frac{\text{L atm}}{\text{mol K}})(673 \text{ K})}{(2.00 \text{ L})} = 0.600 \text{ atm}$$

Finally, we solve for the equilibrium partial pressure in the usual manner:

	P_{HCHO_2}	P_{CO}	P_{H_2O}
I	–	0.401	0.600
C	+x	–x	–x
E	+x	0.401–x	0.600–x

Substituting the above values for equilibrium partial pressures into the mass action expression gives:

$$K_p = \frac{(P_{CO})(P_{H_2O})}{(P_{HCHO_2})} = \frac{(0.401 - x)(0.600 - x)}{x} = 1.6 \times 10^6$$

Because K_p is so large, we assume $x \ll 0.401$ and $x \ll 0.600$. We then solve for $P_{HCHO_2} = x = 1.5 \times 10^{-7}$ atm.

14.79 $K_c = \dfrac{[N_2O][O_2]}{[NO_2][NO]} = 0.914$

Let z = the initial concentration.

	$[NO_2]$	$[NO]$	$[N_2O]$	$[O_2]$
I	z	z	–	–
C	–x	–x	+x	+x
E	z – 2x	z –x	+x	+x

$$K_c = \frac{[x][x]}{[z-x][z-x]} = 0.914$$

Take the square roots of both sides to get

$\dfrac{x}{z-x} = 0.956$ $x = 0.050$ from the data in the problem so

$\dfrac{0.500}{z-0.500} = 0.956$ solving for z we get $z = 0.10$

$\# \text{ moles NO} = \# \text{ moles NO}_2 = \left(\dfrac{0.10 \text{ mol}}{L}\right)(5.00 \text{ L}) = 0.51 \text{ moles}$

Practice Exercises

P15.1 In each case the conjugate base is obtained by removing a proton from the acid:
(a) OH^- (b) I^- (c) NO_2^- (d) $H_2PO_4^-$
(e) HPO_4^{2-} (f) PO_4^{3-} (g) H^- (h) NH_3

P15.2 In each case the conjugate acid is obtained by adding a proton to the base:
(a) H_2O_2 (b) HSO_4^- (c) HCO_3^- (d) HCN
(e) NH_3 (f) NH_4^+ (g) H_3PO_4 (h) $H_2PO_4^-$

P15.3 The Brønsted acids are $H_2PO_4^-(aq)$ and $H_2CO_3(aq)$
The Brønsted bases are $HCO_3^-(aq)$ and $HPO_4^{2-}(aq)$

conjugate pair

$$HCO_3^-(aq) + H_2PO_4^-(aq) \rightleftharpoons H_2CO_3(aq) + HPO_4^{2-}(aq)$$
base   acid   acid   base

conjugate pair

P15.4   conjugate pair

$$PO_4^{3-}(aq) + HC_2H_3O_2(aq) \rightleftharpoons HPO_4^{2-}(aq) + C_2H_3O_2^-(aq)$$
base   acid   acid   base

conjugate pair

P15.5 $HPO_4^{2-}(aq) + OH^-(aq) \rightarrow PO_4^{3-}(aq) + H_2O$; HPO_4^{2-} acting as an acid
$HPO_4^{2-}(aq) + H_3O^+(aq) \rightarrow H_2PO_4^- + H_2O$; HPO_4^{2-} acting as a base

P15.6 The substances on the right because they are the weaker acid and base.

P15.7 (a) HBr is the stronger acid since binary acid strength increases from left to right within a period.
(b) H_2Te is the stronger acid since binary acid strength increases from top to bottom within a group.
(c) CH_3SH since acid strength increases from top to bottom within a group.

P15.8 $HClO_3$ because Cl is more electronegative than Br.

P15.9 (a) HIO_4 (b) H_2SeO_4 (c) H_3AsO_4

P15.10 (a) Fluoride ions have a filled octet of electrons and are likely to behave as Lewis bases, i.e., electron pair donors.
(b) $BeCl_2$ is a likely Lewis acid since it has an incomplete shell. The Be atom has only two valence electrons and it can easily accept a pair of electrons.

P15.11 $K_w = 1.0 \times 10^{-14} = [H^+][OH^-]$

$$[H^+] = \frac{1.0 \times 10^{-14}}{[OH^-]} = \frac{1.0 \times 10^{-14}}{7.8 \times 10^{-6}} = 1.3 \times 10^{-9} \text{ M}$$

Since $[OH^-] > [H^+]$, the solution is basic.

P15.12 $pH = -\log[H^+] = -\log[3.67 \times 10^{-4}] = 3.44$
$pOH = 14.00 - pH = 14.00 - 3.44 = 10.56$
The solution is acidic.

P15.13 $pOH = -\log[OH^-] = -\log[1.47 \times 10^{-9}] = 8.83$
$pH = 14.00 - pOH = 14.00 - 8.83 = 5.17$

P15.14   In general, we have the following relationships between pH and [H⁺]:
$$pH = -\log[H^+] \text{ and } [H^+] = 10^{-pH}$$
   (a) $[H^+] = 10^{-2.90} = 1.3 \times 10^{-3}$ M
           The solution is acidic.
   (b) $[H^+] = 10^{-3.85} = 1.4 \times 10^{-4}$ M
           The solution is acidic.
   (c) $[H^+] = 10^{-10.81} = 1.5 \times 10^{-11}$ M
           The solution is basic.
   (d) $[H^+] = 10^{-4.11} = 7.8 \times 10^{-5}$ M
           The solution is acidic.
   (e) $[H^+] = 10^{-11.61} = 2.5 \times 10^{-12}$ M
           The solution is basic.

P15.15 $[OH^-] = 0.0050$ M
         $pOH = -\log[OH^-] = -\log[0.0050] = 2.30$
         $pH = 14.0 - pOH = 14.00 - 2.30 = 11.70$
         $[H^+] = 10^{-11.70} = 2.0 \times 10^{-12}$ M

Review Problems

15.31   (a)   HF   (b) $N_2H_5^+$ (c) $C_5H_5NH^+$
        (d) HO_2^- (e) H_2CrO_4

15.33   (a)   conjugate pair

$$HNO_3 + N_2H_4 \rightleftharpoons N_2H_5^+ + NO_3^-$$
        acid   base   acid   base

                             conjugate pair

        (b)   conjugate pair

$$N_2H_5^+ + NH_3 \rightleftharpoons NH_4^+ + N_2H_4$$
        acid   base   acid   base

                             conjugate pair

        (c)   conjugate pair

$$H_2PO_4^- + CO_3^{2-} \rightleftharpoons HCO_3^- + HPO_4^{2-}$$
        acid   base   acid   base

                             conjugate pair

        (d)   conjugate pair

$$HIO_3 + HC_2O_4^- \rightleftharpoons H_2C_2O_4 + IO_3^-$$
        acid   base   acid   base

                             conjugate pair

15.35   (a) H_2Se, larger central atom
        (b)   HI, more electronegative atom
        (c) PH_3, larger central atom

15.37   (a) HIO_4, because it has more oxygen atoms

           (b) H_3AsO_4, because it has more oxygen atoms

15.39   (a) H_3PO_4, since P is more electronegative

           (b) HNO_3, because it is a strong acid

           (c) $HClO_4$, because Cl is more electronegative

15.41

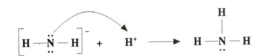

15.43

15.45

15.47   It should be stated from the outset that water at this temperature is neutral by definition, since [H'] = [OH'].

In other words, the self–ionization of water still occurs on a one–to–one mole basis: $H_2O \rightleftharpoons H' + OH'$.

$K_w = 2.4 \times 10^{-14} = [H'][OH']$

Since [H'] = [OH'], we can rewrite the above relationship:

$2.4 \times 10^{-14} = ([H'])^2$, $\therefore$ [H'] = [OH'] = 1.5×10^{-7} M

pH = $-\log[H'] = -\log(1.5 \times 10^{-7}) = 6.82$

pOH = $-\log[OH'] = -\log(1.5 \times 10^{-7}) = 6.82$

pK_w = pH + pOH = 6.82 + 6.82 = 13.64

Alternatively, for the last calculation we can write:

$pK_w = -\log(K_w) = -\log(2.4 \times 10^{-14}) = 13.62$

Water is neutral at this temperature because the concentration of the hydrogen ion is the same as the concentration of the hydroxide ion.

15.49   At 25 °C, $K_w = 1.0 \times 10^{-14} = [H'][OH']$. Let x = [H'], for each of the following:

           (a) $x(0.0024) = 1.0 \times 10^{-14}$

                    [H'] = $(1.0 \times 10^{-14}) \div (0.0024) = 4.2 \times 10^{-12}$ M

           (b) $x(1.4 \times 10^{-5}) = 1.0 \times 10^{-14}$

                    [H'] = $(1.0 \times 10^{-14}) \div (1.4 \times 10^{-5}) = 7.1 \times 10^{-10}$ M

           (c) $x(5.6 \times 10^{-9}) = 1.0 \times 10^{-14}$

                    [H'] = $(1.0 \times 10^{-14}) \div (5.6 \times 10^{-9}) = 1.8 \times 10^{-6}$ M

           (d) $x(4.2 \times 10^{-13}) = 1.0 \times 10^{-14}$

                    [H'] = $(1.0 \times 10^{-14}) \div (4.2 \times 10^{-13}) = 2.4 \times 10^{-2}$ M

15.51 $pH = -\log[H^+]$

       $[H^+] = 4.2 \times 10^{-12}$ M $pH = 11.38$

       $[H^+] = 7.1 \times 10^{-10}$ M $pH = 9.15$

       $[H^+] = 1.8 \times 10^{-6}$ M $pH = 5.74$

       $[H^+] = 2.4 \times 10^{-2}$ M $pH = 1.62$

15.53 $pH = -\log[H^+] = -\log(1.9 \times 10^{-5}) = 4.72$

15.55 $[H^+] = 10^{-pH}$ and $[OH^-] = 10^{-pOH}$

       At 25 °C, $pH + pOH = 14.00$

    (a) $[H^+] = 10^{-pH} = 10^{-3.14} = 7.2 \times 10^{-4}$ M

            $pOH = 14.00 - pH = 14.00 - 3.14 = 10.86$

            $[OH^-] = 10^{-pOH} = 10^{-10.86} = 1.4 \times 10^{-11}$ M

    (b) $[H^+] = 10^{-pH} = 10^{-2.78} = 1.7 \times 10^{-3}$ M

            $pOH = 14.00 - pH = 14.00 - 2.78 = 11.22$

            $[OH^-] = 10^{-pOH} = 10^{-11.22} = 6.0 \times 10^{-12}$ M

    (c) $[H^+] = 10^{-pH} = 10^{-9.25} = 5.6 \times 10^{-10}$ M

            $pOH = 14.00 - pH = 14.00 - 9.25 = 4.75$

            $[OH^-] = 10^{-pOH} = 10^{-4.75} = 1.8 \times 10^{-5}$ M

    (d) $[H^+] = 10^{-pH} = 10^{-13.24} = 5.8 \times 10^{-14}$ M

            $pOH = 14.00 - pH = 14.00 - 13.24 = 0.76$

            $[OH^-] = 10^{-pOH} = 10^{-0.76} = 1.7 \times 10^{-1}$ M

    (e) $[H^+] = 10^{-pH} = 10^{-5.70} = 2.0 \times 10^{-6}$ M

            $pOH = 14.00 - pH = 14.00 - 5.70 = 8.30$

            $[OH^-] = 10^{-pOH} = 10^{-8.30} = 5.0 \times 10^{-9}$ M

15.57   HCl is a strong acid so $[H^+] = [HCl] = 0.010$ M

       $pH = -\log[H^+] = -\log(0.010) = 2.00$

15.59 $$M\,OH^- = \frac{\#\,\text{moles}\,OH^-}{\#\,\text{L solution}} = \left(\frac{6.0\,\text{g NaOH}}{1.00\,\text{L solution}}\right)\left(\frac{1\,\text{mole NaOH}}{40.0\,\text{g NaOH}}\right)\left(\frac{1\,\text{mole}\,OH^-}{1\,\text{mole NaOH}}\right)$$

            $= 0.15\,M\,OH^-$

       $pOH = -\log[OH^-] = -\log(0.15) = 0.82$

       $pH = 14.00 - pOH = 14.00 - 0.82 = 13.18$

15.61 $pOH = 14.00 - pH = 14.00 - 11.60 = 2.40$

       $[OH^-] = 10^{-pOH} = 10^{-2.40} = 4.0 \times 10^{-3}$ M

       $$[Ca(OH)_2] = \left(\frac{4.0 \times 10^{-3}\,\text{mol}\,OH^-}{1\,\text{L solution}}\right)\left(\frac{1\,\text{mol}\,Ca(OH)_2}{2\,\text{mol}\,OH^-}\right)$$

            $= 2.0 \times 10^{-3}\,M\,Ca(OH)_2$

Additional Exercises

15.63   Acid: $(CH_3)_2NH_2^+$ Base: $(CH_3)_2N^-$

15.65   (a)   In H_2O_2, the oxygen atom helps to stabilize the HO_2^- ion making it easier for H_2O_2 to lose a proton than can H_2O.
         (b)   acidic

15.67   The equilibrium lies to the right since reactions favor the weaker acid and base.

15.70

$$\# \text{ moles } H^+ = (38.0 \text{ mL})\left(\frac{0.00200 \text{ moles}}{1000 \text{ mL}}\right) = 7.60 \times 10^{-5} \text{ moles } H^+$$

$$\# \text{ moles } OH^- = (40.0 \text{ mL})\left(\frac{0.0018 \text{ moles}}{1000 \text{ mL}}\right) = 7.20 \times 10^{-5} \text{ moles } OH^-$$

$$\text{excess } H^+ = 0.40 \times 10^{-5} \text{ moles } H^+$$

$$[H^+] = \frac{0.40 \times 10^{-5} \text{ moles}}{(38 \text{ mL} + 40 \text{ mL})\left(\frac{1 \text{ L}}{1000 \text{ mL}}\right)} = 5.12 \times 10^{-5} \text{ M}$$

$$pH = 4.29$$

15.71   The total $[H^+]$ is from the HCl and from the dissociation of H_2O. Since HCl is a strong acid, it will contribute 1.0×10^{-7} mol of H^+ per liter of solution. We need to use the equilibrium expression to determine the amount of H^+ contributed by the water.
         $$K_w = [H^+][OH^-] = (1.0 \times 10^{-7} + x)(x) = 1.0 \times 10^{-14}$$
         Solving a quadratic equation we see that $x = 6 \times 10^{-8} = [OH^-] = [H^+]$ from water dissociation.
         So, $[H^+]_{total} = 1.0 \times 10^{-7} + 6 \times 10^{-8} = 1.6 \times 10^{-7}$ and $pH = 6.80$.

Practice Exercises

P16.1　(a)　$HCHO_2 + H_2O \rightleftharpoons H_3O^+ + CHO_2^-$

$$K_a = \frac{[H_3O^+][CHO_2^-]}{[HCHO_2]}$$

(b)　$(CH_3)_2NH_2^+ + H_2O \rightleftharpoons H_3O^+ + (CH_3)_2NH$

$$K_a = \frac{[H_3O^+][CH_3NH]}{[(CH_3)_2NH_2^+]}$$

(c)　$H_2PO_4^- + H_2O \rightleftharpoons H_3O^+ + HPO_4^{2-}$

$$K_a = \frac{[H_3O^+][HPO_4^{2-}]}{[H_2PO_4^-]}$$

P16.2　The acid with the smaller pKa, HA, is the strongest acid. Since pKa = – log Ka
Ka = 10^{-pKa}
For HA: Ka = $10^{-3.16}$ = 6.9 x 10^{-4}
For HB: Ka = $10^{-4.14}$ = 7.2 x 10^{-5}

P16.3　(a)　$(CH_3)_3N + H_2O \rightleftharpoons (CH_3)_3NH^+ + OH^-$

$$K_b = \frac{[(CH_3)_3NH^+][OH^-]}{[(CH_3)_3N]}$$

(b)　$SO_3^{2-} + H_2O \rightleftharpoons HSO_3^- + OH^-$

$$K_b = \frac{[HSO_3^-][OH^-]}{[SO_3^{2-}]}$$

(c)　$NH_2OH + H_2O \rightleftharpoons NH_2OH_2^+ + OH^-$

$$K_b = \frac{[NH_3OH_2^+][OH^-]}{[NH_3OH]}$$

P16.4　For conjugate acid base pairs, $K_a \times K_b = K_w$.
$K_b = K_w \div K_a$ = 1.0 X 10^{-14} ÷ 1.8 X 10^{-4} = 5.6 X 10^{-11}.

P16.5　$HBu \rightleftharpoons H^+ + Bu^-$

$$K_a = \frac{[H^+][Bu^-]}{[HBu]}$$

	[HBu]	[H⁺]	[Bu⁻]
I	0.01000	–	–
C	–x	+x	+x
E	0.01000–x	+x	+x

We know that the acid is 4.0% ionized so x = 0.01000 M × 0.040 = 0.00040 M. Therefore, our equilibrium concentrations are [H⁺] = [Bu⁻] = 0.00040 M, and [HBu] = 0.01000 M – 0.00040 M = 0.00960 M.

Substituting these values into the mass action expression gives:

$$K_a = \frac{(0.00040)(0.00040)}{0.00960} = 1.7 \times 10^{-5}$$
$$pK_a = -\log(K_a) = -\log(1.7 \times 10^{-5}) = 4.78$$

P16.6   We will use the symbol Mor and HMor⁺ for the base and its conjugate acid respectively:

$$Mor + H_2O \rightleftharpoons HMor^\cdot + OH^-, \qquad K_b = \frac{[HMor^+][OH^-]}{[Mor]}$$

	[Mor]	[HMor⁺]	[OH⁻]
I	0.010	—	—
C	−x	+x	+x
E	0.010−x	+x	+x

At equilibrium, the pH = 10.10 and the pOH = 14.00 − 10.10 = 3.90.
The $[OH^-] = 10^{-pOH} = 10^{-3.90} = 1.3 \times 10^{-4}$ M = x.

Substituting these values into the mass action expression gives:

$$K_b = \frac{(1.3 \times 10^{-4})(1.3 \times 10^{-4})}{0.010 - 1.3 \times 10^{-4}} = 1.6 \times 10^{-6}$$
$$pK_b = -\log(K_b) = -\log(1.6 \times 10^{-6}) = 5.79$$

P16.7
$$HC_2H_6NO_2 \rightleftharpoons H^\cdot + C_2H_6NO_2^-$$
$$K_a = \frac{[H^+][C_2H_6NO_2^-]}{[HC_2H_6NO_2]} = 1.4 \times 10^{-5}$$

	[HC₂H₆NO₂]	[H⁺]	[C₂H₆NO₂⁻]
I	0.050	—	—
C	−x	+x	+x
E	0.050−x	+x	+x

Assume that x << 0.050 and substitue the equilibrium values into the mass action expression to get:
$$K_a = \frac{[x][x]}{[0.050]} = 1.4 \times 10^{-5}$$
Solving for x we determine that x = 8.4×10^{-4} M = [H⁺].
pH = −log[H⁺] = −log(8.4×10^{-4}) = 3.08

P16.8
$$C_5H_5N + H_2O \rightleftharpoons C_5H_5NH^\cdot + OH^-$$
$$K_b = \frac{[C_5H_5NH^+][OH^-]}{[C_5H_5N]} = 1.5 \times 10^{-9}$$

	[C₅H₅N]	[C₅H₅NH⁺]	[OH⁻]
I	0.010	—	—
C	−x	+x	+x
E	0.010−x	+x	+x

Assume that x << 0.010 and substitue the equilibrium values into the mass action expression to get:
$$K_b = \frac{[x][x]}{[0.010]} = 1.5 \times 10^{-9}$$

Solving for x we determine that x = 3.9 X 10^{-6} M = [OH⁻].
pOH = –log[OH⁻] = –log(3.9 X 10^{-6}) = 5.41
pH = 14.00 – pOH = 8.59

P16.9   We will use the notation Hphenol and phenol for the acid and its conjugate base.

Hphenol $\rightleftharpoons$ H⁺ + phenol, $K_a = \dfrac{[H^+][phenol]}{[Hphenol]} = 1.3 \times 10^{-10}$

	[Hphenol]	[H⁺]	[phenol]
I	0.15	–	–
C	–x	+x	+x
E	0.15–x	+x	+x

If we assume that x << 0.15, a good assumption based upon the size of K_a, we can substitute the equilibrium values in to the mass action expression to get:

$$K_a = \frac{(x)(x)}{0.15} = 1.3 \times 10^{-10}$$

Solving gives x = 4.4 X 10^{-6} M = [H⁺]
pH = –log[H⁺] = –log(4.4 X 10^{-6}) = 5.36

P16.10 (a)   Sodium ions are the salt of a strong base, NaOH, and are therefore neutral. The nitrite ion is the salt of a weak acid, HNO_2 (nitrous acid), and is therefore basic. Consequently, a solution of $NaNO_2$ will be basic.

(b)   Potassium ions are the salt of a strong base, KOH, and are therfore neutral. The chloride ion is the salt of a strong acid, HCl, and is therefore neutral as well. The KCl solution will be neutral since both salts are neutral.

(c)   The ammonium ion is the salt of a weak base, NH_3, and is consequently acidic. Bromide ions are the salt of a strong acid, HBr, and are thus neutral. The NH_4Br solution will be acidic as it combines an acidic salt and a neutral salt.

P16.11 The sodium ion is neutral since it is the salt of the strong base, NaOH. The nitrite ion is basic since it is the salt of nitrous acid, HNO_2, a weak acid. The equilibrium we are interested in for this problem is: NO_2^- + $H_2O \rightleftharpoons HNO_2$ + OH⁻.

$$K_b = \frac{[HNO_2][OH^-]}{[NO_2^-]}$$

In order to determine the value for K_b recall that $K_a \times K_b = K_w$. We can look for the value of K_a for HNO_2 K_b = 1.0 X 10^{-14} ÷ 7.1 X 10^{-4} = 1.4 X 10^{-11}.

	[NO₂⁻]	[HNO₂]	[OH⁻]
I	0.10	–	–
C	–x	+x	+x
E	0.10–x	+x	+x

Assume that x << 0.10 and substitue the equilibrium values into the mass action expression to get:

$$K_b = \frac{[x][x]}{[0.10]} = 1.4 \times 10^{-11}$$

Solving we determine that x = 1.2 X 10^{-6} M = [OH⁻].
pOH = –log[OH⁻] = –log(1.2 X 10^{-6}) = 5.93
pH = 14.00 – pOH = 8.07

P16.12 As previously determined, a solution of NH_4Br will be acidic since NH_4^+ is the salt of a weak base and Br^- is the salt of a strong acid. As in the previous Practice Exercise, we need to determine the value for the dissociation constant using the relationship $K_a \times K_b = K_w$ and the value of K_b for NH_3 as listed in Table 16.5. $K_a = 1.0 \times 10^{-14} \div 1.8 \times 10^{-5} = 5.6 \times 10^{-10}$. The equilibrium reaction is

$$NH_4^+ \rightleftharpoons NH_3 + H^+$$

	$[NH_4^+]$	$[NH_3]$	$[H^+]$
I	0.10	–	–
C	–x	+x	+x
E	0.10–x	+x	+x

Assume that x << 0.10 and substitue the equilibrium values into the mass action expression to get:

$$K_a = \frac{[x][x]}{[0.10]} = 5.6 \times 10^{-10}$$

Solving we determine that $x = 7.5 \times 10^{-6}\ M = [H^+]$.
$pH = -\log[H^+] = -\log(7.5 \times 10^{-6}) = 5.13$

P16.13 Since the ammonium ion is the salt of a weak base, NH_3, it is acidic. The cyanide ion is the salt of a weak acid, HCN, so it is basic. In order to determine if the solution is acidic or basic, we need to determine the relative strength of the two components. Use the relationship $K_a \times K_b = K_w$ in order to determine the dissociation constants for the cyanide ion and the ammonium ion.

$$K_a(NH_4^+) = K_w \div K_b(NH_3) = 1.0 \times 10^{-14} \div 1.8 \times 10^{-5} = 5.6 \times 10^{-10}$$
$$K_b(CN^-) = K_w \div K_a(HCN) = 1.0 \times 10^{-14} \div 6.2 \times 10^{-14} = 1.6 \times 10^{-5}$$

Since the $K_b(CN^-)$ is larger than the $K_a(NH_4^+)$ the NH_4CN solution will be basic.

P16.14 $$(CH_3)_2NH + H_2O \rightleftharpoons (CH_3)_2NH_2^+ + OH^-$$

$$K_b = \frac{[(CH_3)_2NH_2^+][OH^-]}{[(CH_3)_2NH]} = 9.6 \times 10^{-4}$$

	$[(CH_3)_2NH]$	$[(CH_3)_2NH_2^+]$	$[OH^-]$
I	0.0010	–	–
C	–x	+x	+x
E	0.0010–x	+x	+x

We cannot neglect x in this calculation due to the large size of the dissociation constant. Consequently, we will have to solve the quadratic equation. Substituting the equilibrium values into the mass action expression gives:

$$K_b = \frac{[x][x]}{[0.0010 - x]} = 9.6 \times 10^{-4}$$

Rearranging and collecting terms on one side of the equal sign gives:
$$x^2 + 9.6 \times 10^{-4} x - 9.6 \times 10^{-7} = 0$$
Using the quadratic equation
$x = 6.1 \times 10^{-4}\ M = [OH^-]$, pOH = 3.21 and the pH = 10.79.

P16.15 The equation is:
$$C_2H_3O_2^- + H_2O \rightleftharpoons HC_2H_3O_2 + OH^-$$
Start by determining Kb for acetate ion using $K_w = K_aK_b$
$K_b = K_w/K_a = 1.0 \times 10^{-14}/1.8 \times 10^{-5} = 5.6 \times 10^{-10}$

$$K_b = \frac{[HC_2H_3O_2][OH^-]}{[C_2H_3O_2^-]}$$

	$[C_2H_3O_2^-]$	$[HC_2H_3O_2]$	$[OH^-]$
I	0.11	0.090	–
C	–x	+x	+x
E	0.11 – x	0.090 +x	+x

$$K_b = \frac{[x][0.090 + x]}{[0.11 - x]}$$

assume x << 0.090 and solve for x

$$x = [OH^-] = 6.8 \times 10^{-10}$$

$$pOH = 9.16$$

$$pH = 14.00 - 9.16 = 4.84$$

P16.16 We will use the relationship $K_a \times K_b = K_w$.

$$K_a(NH_4^+) = K_w \div K_b(NH_3) = 1.0 \times 10^{-14} \div 1.8 \times 10^{-5} = 5.6 \times 10^{-10}$$

The equation for the acid dissociation is $NH_4^+ \rightleftharpoons H^+ + NH_3$. The dissociation constant is written:

$$K_a = \frac{[H^+][NH_3]}{[NH_4^+]} = 5.6 \times 10^{-10} = \frac{[H^+](0.12)}{(0.095)}$$

Solving gives $[H^+] = 4.4 \times 10^{-10}$ M

$$pH = -\log[H^+] = -\log(4.4 \times 10^{-10}) = 9.36$$

P16.17 Yes, formic acid and sodium formate would make a good buffer solution since $pK_a = 3.74$ and the desired pH is within one pH unit of this value.

Using Equation 16.16; $\left[H^+\right] = K_a \times \dfrac{mol\, HCHO_{2\,initial}}{mol\, CHO_2^-{}_{initial}}$

Rearranging and substituting the known values we get;

$$\frac{mol\, HCHO_{2\,initial}}{mol\, CHO_2^-{}_{initial}} = \frac{\left[H^+\right]}{K_a} = \frac{1.3 \times 10^{-4}}{1.8 \times 10^{-4}} = 0.70$$

P16.18 Using the data provided in Example 16.20, the initial pH of this buffer is 4.74. When OH^- is added to the solution, it will react with the H^+ present from the dissociation of the acetic acid. The acetic acid in solution dissociates further to maintain the equilibrium and any unreacted hydroxide will react with H^+ as it is produced. Eventually, the hydroxide will be completely reacted. The net change that occurs is a reduction in the amount of acetic acid in solution and an equivalent increase in the amount of acetate ion in solution. We started with 1 mol of acetic acid and 1 mol of acetate ion. The addition of 0.11 mol OH^- will reduce the amount of acetic acid to 0.89 mol and increase the amount of acetate ion to 1.11 mol. Substituting these amounts into the mass action expression gives:

$$K_a = \frac{[H^+](1.11)}{(0.89)} = 1.8 \times 10^{-5}$$

$$[H^+] = 1.4 \times 10^{-5}\, M$$

pH = 4.84.

P16.19 [H'] is determined by the first protic equilibrium:

$$H_2C_6H_6O_6 \rightleftharpoons H^+ + HC_6H_6O_6^-$$

The mass action expression is:

$$K_{a1} = 6.8 \times 10^{-5} = x^2/0.10$$

$$x = [H^+] = 2.6 \times 10^{-3}\ M$$

$$pH = -\log(2.6 \times 10^{-3}) = 2.58$$

The concentration of the anion, $[HC_6H_6O_6^-]$, is given almost entirely by the second ionization equilibrium:

$$HC_6H_6O_6^- \rightleftharpoons H^+ + C_6H_6O_6^{2-}$$

for which the mass action expression is:

$$K_{a2} = \frac{[H^+][C_6H_6O_6^{2-}]}{[HC_6H_6O_6^-]} = 2.7 \times 10^{-12}$$

We have used the value for K_{a2} from Table 17.1. Using the value of x from the first step above gives:

$$2.7 \times 10^{-12} = \frac{(2.6 \times 10^{-3})[C_6H_6O_6^{2-}]}{(2.6 \times 10^{-3})}$$

$$[C_6H_6O_6^{2-}] = 2.7 \times 10^{-12}$$

P16.20 The equilibrium we are interested in for this problem is

$$SO_3^{2-}(aq) + H_2O(\ell) \rightleftharpoons HSO_3^-(aq) + OH^-(aq)$$

$$K_b = K_w/K_{a2} = 1.0 \times 10^{-14} / 6.6 \times 10^{-8} = 1.5 \times 10^{-7}$$

$$K_b = 1.5 \times 10^{-7} = \frac{[HSO_3^-][OH^-]}{[SO_3^{2-}]}$$

	$[SO_3^{2-}]$	$[HSO_3^-]$	$[OH^-]$
I	0.20	–	–
C	–x	+x	+x
E	0.20–x	+x	+x

Substituting these values into the mass action expression gives:

$$K_b = 1.5 \times 10^{-7} = \frac{(x)(x)}{0.20 - x}$$

Assume that x << 0.20 and solving gives x = 1.7 X 10⁻⁴.

$$x = [OH^-] = 1.7 \times 10^{-4}\ M$$

$$pOH = -\log(1.7 \times 10^{-4}) = 3.76$$

$$pH = 14.00 - pOH = 14.00 - 3.76 = 10.24$$

P16.21 For weak polyprotic acids, the concentration of the polyvalent ions is equal to the volume of K_{an} where n is the valency. By analogy, the concentration of H_2SO_3 in 0.010 M Na_2SO_3 will be equal to K_{b2} for SO_3^{2-}.

P16.22 $$HCHO_2 + H_2O \rightleftharpoons H_3O^+ + CHO_2^-$$

$$Ka = \frac{[H_3O^+][CHO_2^-]}{[HCHO_2]} = 1.8 \times 10^{-4}$$

a)

	[HCHO$_2$]	[H$_3$O$^+$]	[CHO$_2^-$]
I	0.100	–	–
C	–x	+x	+x
E	0.100–x	+x	+x

Assume x << 0.100. Solving we get x = [H$_3$O$^+$] = 4.2 X 10^{-3}. The pH is 2.37.

b) [HCHO$_2$] = [CHO$_2^-$] so [H$_3$O$^+$] = Ka = 1.8 X 10^{-4} and the pH = 3.74.

c)

$$\text{\# moles base added} = (15.0\text{ mL})\left(\frac{0.100\text{ mol}}{1000\text{ mL}}\right) = 1.50 \times 10^{-3}$$

$$\text{\# moles acid initially} = (20.0\text{ mL})\left(\frac{0.10\text{ mol}}{1000\text{ mL}}\right) = 2.00 \times 10^{-3}$$

$$\text{excess acid} = 2.00 \times 10^{-3} - 1.50 \times 10^{-3} = 0.50 \times 10^{-3}\text{ moles acid}$$

$$[\text{acid}] = \frac{0.50 \times 10^{-3}\text{ moles}}{(35\text{ mL})\left(\frac{1\text{ L}}{1000\text{ mL}}\right)} = 1.43 \times 10^{-2}\text{ M}$$

$$[\text{base}] = \frac{1.50 \times 10^{-3}\text{ moles}}{(35\text{ mL})\left(\frac{1\text{ L}}{1000\text{ mL}}\right)} = 4.29 \times 10^{-2}\text{ M}$$

Substituting into the equilibrium expression and solving we get [H$_3$O$^+$] = 6.00 X 10^{-5} and the pH = 4.22.

(d) We now have a solution of formate ion with a concentration of 0.0500 M. We need Kb for formate ion: Kb = Kw/Ka = 5.6 x 10^{-11}. If we set up the equilibrium problem and solve we get: [OH$^-$] = 1.7 x 10^{-6}. The pOH = 5.78 and the pH = 8.22.

P16.23

$$\text{\# moles base added} = (30.0\text{ mL})\left(\frac{0.15\text{ mol}}{1000\text{ mL}}\right) = 4.50 \times 10^{-3}$$

$$\text{\# moles acid initially} = (50.0\text{ mL})\left(\frac{0.20\text{ mol}}{1000\text{ mL}}\right) = 1.00 \times 10^{-2}$$

$$\text{excess acid} = 1.00 \times 10^{-2} - 4.50 \times 10^{-3} = 5.50 \times 10^{-3}\text{ moles acid}$$

$$[\text{acid}] = \frac{5.50 \times 10^{-3}\text{ moles}}{(80\text{ mL})\left(\frac{1\text{ L}}{1000\text{ mL}}\right)} = 6.88 \times 10^{-2}\text{ M}$$

$$[\text{base}] = \frac{4.50 \times 10^{-3}\text{ moles}}{(80\text{ mL})\left(\frac{1\text{ L}}{1000\text{ mL}}\right)} = 5.63 \times 10^{-2}\text{ M}$$

$$HCHO_2 + H_2O \rightleftharpoons H_3O^+ + CHO_2^-$$

$$Ka = \frac{[H_3O^+][CHO_2^-]}{[HCHO_2]}$$

	[HCHO$_2$]	[H$_3$O$^+$]	[CHO$_2^-$]
I	6.88 x 10^{-2}	–	5.63 x 10^{-2}
C	–x	+x	+x
E	6.88 x 10^{-2}–x	+x	5.63 x 10^{-2}+x

Assume x << 5.63 x 10^{-2}

$$Ka = \frac{[x][5.63 \times 10^{-2}]}{[6.88 \times 10^{-2}]}, \quad x = 2.20 \times 10^{-2} = [H_3O^+]$$
$$pH = 3.66$$

Review Problems

16.32　At 25 °C, $K_a K_b = K_w$
　　　$K_b = K_w/K_a = 1.0 \times 10^{-14} \div 6.8 \times 10^{-4} = 1.5 \times 10^{-11}$

16.34　At 25 °C, $K_a K_b = K_w$
　　　$K_b = K_w/K_a = 1.0 \times 10^{-14} \div 1.8 \times 10^{-12} = 5.6 \times 10^{-3}$

16.36　At 25 °C, $K_a K_b = K_w$
　　　$K_b = K_w/K_a = 1.0 \times 10^{-14} \div 1.4 \times 10^{-4} = 7.1 \times 10^{-11}$

16.38　$HIO_4 \rightleftharpoons H^+ + IO_4^-$
　　　$K_a = \dfrac{\left[H^+\right]\left[IO_4^-\right]}{\left[HIO_4\right]}$

	[HIO$_4$]	[H$^+$]	[IO$_4^-$]
I	0.10	–	–
C	–x	+x	+x
E	0.10–x	+x	+x

We know that at equilibrium [H$^+$] = 0.038 M = x. The equilibrium concentrations of the other components of the mixture are:
　　　[HIO$_4$] = 0.10 – x = 0.06 M and [IO$_4^-$] = x = 0.038 M.

Substituting the above values for equilibrium concentrations into the mass action expression gives:

$$K_a = \frac{(0.038)(0.038)}{0.06} = 2 \times 10^{-2}$$
$$pK_a = -\log(K_a) = -\log(2 \times 10^{-2}) = 1.7$$

16.40　pOH = 14.00 – pH = 14.00 – 11.86 = 2.14
　　　[OH$^-$] = 10^{-pOH} = 10$^{-2.14}$ = 7.2 X 10^{-3} M

$$CH_3CH_2NH_2 + H_2O \rightleftharpoons CH_3CH_2NH_3^+ + OH^-$$
$$K_b = \frac{\left[CH_3CH_2NH_3^+\right]\left[OH^-\right]}{\left[CH_3CH_2NH_2\right]}$$

	$[CH_3CH_2NH_2]$	$[CH_3CH_2NH_3^+]$	$[OH^-]$
I	0.10	–	–
C	–x	+x	+x
E	0.10–x	+x	+x

In the equilibrium analysis, the value of x is, therefore, equal to 7.2×10^{-3} M. Therefore, our equilibrium concentrations are $[CH_3CH_2NH_3^+] = [OH^-] = 7.2 \times 10^{-3}$ M, and $[CH_3CH_2NH_2] = 0.10$ M – 7.2×10^{-3} M = 0.09 M.

Substituting these values into the mass action expression gives:

$$K_b = \frac{\left(7.2 \times 10^{-3}\right)\left(7.2 \times 10^{-3}\right)}{0.09} = 6 \times 10^{-4}$$

$$pK_b = -\log(K_b) = -\log(6 \times 10^{-4}) = 3.2$$

16.42 $HC_3H_5O_2 + H_2O \rightleftharpoons H_3O^+ + C_3H_5O_2^-$

$$K_a = \frac{[H_3O^+][C_3H_5O_2^-]}{[HC_3H_5O_2]} = 1.4 \times 10^{-4}$$

	$[HC_3H_5O_2]$	$[H_3O^+]$	$[C_3H_5O_2^-]$
I	0.15	–	–
C	–x	+x	+x
E	0.15 –x	+x	+x

Assume x << 0.15

$$K_a = \frac{[x][x]}{[0.15]} = 1.4 \times 10^{-4} \quad x = 4.6 \times 10^{-3} = [H_3O^+]$$

$$pH = 2.34$$

$[HC_3H_5O_2] = 0.15$
$[H^+] = 4.6 \times 10^{-3}$
$[C_3H_5O_2^-] = 4.6 \times 10^{-3}$

16.44 $HN_3 + H_2O \rightleftharpoons H_3O^+ + N_3^-$

$$K_a = \frac{[H_3O^+][N_3^-]}{[HN_3]} = 1.8 \times 10^{-5}$$

	$[HN_3]$	$[H_3O^+]$	$[N_3^-]$
I	0.15	–	–
C	–x	+x	+x
E	0.15 –x	+x	+x

Assume x << 0.15

$$K_a = \frac{[x][x]}{[0.15]} = 1.8 \times 10^{-5} \quad x = 1.6 \times 10^{-3} = [H_3O^+]$$
$$pH = 2.78$$

16.46 $K_b = 10^{-pK_b} = 10^{-5.79} = 1.6 \times 10^{-6}$

$Cod + H_2O \rightleftharpoons HCod^+ + OH^-$

$$K_b = \frac{[HCod^+][OH^-]}{[Cod]} = 1.6 \times 10^{-6}$$

	[Cod]	[HCod$^+$]	[OH$^-$]
I	0.020	–	–
C	–x	+x	+x
E	0.020–x	+x	+x

Substituting these values into the mass action expression gives:

$$K_b = \frac{(x)(x)}{0.020 - x} = 1.6 \times 10^{-6}$$

If we assume that x << 0.020 we get; $x^2 = 3.2 \times 10^{-8}$,
x = 1.8 × 10^{-4} M = [OH$^-$]
pOH = –log[OH$^-$] = –log(1.8 × 10^{-4}) = 3.74
pH = 14.00 – pOH = 14.00 – 3.74 = 10.26

16.48 $[H^+] = 10^{-pH} = 10^{-2.54} = 2.9 \times 10^{-3}$ M

$HC_2H_3O_2 \rightleftharpoons H^+ + C_2H_3O_2^-$

$$K_a = \frac{[H^+][C_2H_3O_2^-]}{[HC_2H_3O_2]} = 1.8 \times 10^{-5}$$

	[HC$_2$H$_3$O$_2$]	[H$^+$]	[C$_2$H$_3$O$_2^-$]
I	Z	–	–
C	–x	+x	+x
E	Z–x	+x	+x

Substituting these values into the mass action expression gives:

$$K_a = \frac{(x)(x)}{Z - x} = 1.8 \times 10^{-5}$$

Assuming x << Z and knowing that x = 2.9 × 10^{-3} M, we can solve for Z and find Z = 0.47. The initial concentration of HC$_2$H$_3$O$_2$ is 0.47 M.

16.50 NaCN will be basic in solution since CN$^-$ is a basic ion and Na$^+$ is a neutral ion.
$CN^- + H_2O \rightleftharpoons HCN + OH^-$
For HCN, $K_a = 6.2 \times 10^{-10}$, we need K_b for CN$^-$;
$K_b = K_w/K_a = (1.0 \times 10^{-14}) \div (6.2 \times 10^{-10}) = 1.6 \times 10^{-5}$
$$K_b = \frac{[HCN][OH^-]}{[CN^-]} = 1.6 \times 10^{-5}$$

	[CN⁻]	[HCN]	[OH⁻]
I	0.20	–	–
C	–x	+x	+x
E	0.20–x	+x	+x

Substituting these values into the mass action expression gives:

$$K_b = \frac{(x)(x)}{0.20 - x} = 1.6 \times 10^{-5}$$

Assuming that x << 0.20 we can solve for x an determine;
x = 1.8 X 10⁻³ M = [OH⁻]
pOH = –log[OH⁻] = –log(1.8 X 10⁻³) = 2.74
pH = 14.00 – pOH = 14.00 – 2.74 = 11.26

16.52   A solution of CH₃NH₃Cl will be acidic since the Cl⁻ ion is neutral and the CH₃NH₃⁺ ion is acidic.
CH₃NH₃⁺ ⇌ H⁺ + CH₃NH₂
For CH₃NH₂, K_b = 4.4 X 10⁻⁴. We need K_a for CH₃NH₃⁺;
K_a = K_w/K_b = (1.0 X 10⁻¹⁴) ÷ (4.4 X 10⁻⁴) = 2.3 X 10⁻¹¹

$$K_a = \frac{[H^+][CH_3NH_2]}{[CH_3NH_3^+]} = 2.3 \times 10^{-11}$$

	[CH₃NH₃⁺]	[H⁺]	[CH₃NH₂]
I	0.15	–	–
C	–x	+x	+x
E	0.15–x	+x	+x

Substituting these values into the mass action expression gives:

$$K_a = \frac{(x)(x)}{0.15 - x} = 2.3 \times 10^{-11}$$

Assuming that x << 0.15 we can solve for x an determine;
x = 1.9 X 10⁻⁶ M = [H₃O⁺]
pH = –log[H₃O⁺] = –log(1.9 X 10⁻⁶) = 5.72

16.54   The reaction for this problem is:
H–Mor⁺ ⇌ H⁺ + Mor
We know that pK_a + pK_b = pK_w = 14.00.

So, pK_a = 14.00 – pK_b = 14.00 – 6.13 = 7.87, and K_a = 10⁻ᵖᴷᵃ = 1.3 X 10⁻⁸.

$$K_a = \frac{[H^+][Mor]}{[H-Mor^+]} = 1.3 \times 10^{-8}$$

	[H–Mor⁺]	[H⁺]	[Mor]
I	0.20	–	–
C	–x	+x	+x
E	0.20–x	+x	+x

Substituting these values into the mass action expression gives:

$$K_a = \frac{(x)(x)}{0.20 - x} = 1.3 \times 10^{-8}$$

Assuming that x << 0.20 we can solve for x an determine;
x = 5.1 X 10^{-5} M = [H$^+$]
pH = –log[H$^+$] = –log(5.1 X 10^{-5}) = 4.29

16.56   Let HNic symbolize the nicotinic acid
Nic$^-$ + H$_2$O $\rightleftharpoons$ HNic + OH$^-$

$$K_b = \frac{[HNic][OH^-]}{[Nic^-]}$$

	[Nic$^-$]	[HNic]	[OH$^-$]
I	0.18	–	–
C	–x	+x	+x
E	0.18–x	+x	+x

Assume x << 0.18

$$K_b = \frac{(x)(x)}{0.18}$$ we know x = [OH$^-$]

we know pH = 9.05 so pOH = 4.95

and [OH$^-$] = 10^{-pOH} = 1.1 x 10^{-5} = x

$$Now\ K_b = \frac{(1.1 \times 10^{-5})^2}{0.18} = 7.0 \times 10^{-10}$$

16.60   OCl$^-$ + H$_2$O $\rightleftharpoons$ HOCl + OH$^-$

$$K_b = \frac{[HOCl][OH^-]}{[OCl^-]} = \frac{K_w}{K_a} = \frac{1.0 \times 10^{-14}}{3.0 \times 10^{-8}} = 3.3 \times 10^{-7}$$

$$[OCl^-] = \left(\frac{5.0\,g\,NaOCl}{100\,g\,solution}\right)\left(\frac{1\,mol\,NaOCl}{74.5\,g\,NaOCl}\right)\left(\frac{1.0\,g}{1\,mL}\right)\left(\frac{1000\,mL}{1\,L}\right)$$
$$= 0.67\,M$$

	[OCl$^-$]	[HOCl]	[OH$^-$]
I	0.67	–	–
C	–x	+x	+x
E	0.67 –x	+x	+x

Assume that x << 0.67

$$K_b = \frac{(x)(x)}{0.67} = 3.3 \times 10^{-7} \quad x = 4.7 \times 10^{-4} = [OH^-]$$
pOH = 3.33
pH = 10.67

16.62 $HF \rightleftharpoons H^+ + F^-$

$$K_a = \frac{[H^+][F^-]}{[HF]} = 6.8 \times 10^{-4}$$

	[HF]	[H$^+$]	[F$^-$]
I	0.15	—	—
C	−x	+x	+x
E	0.15−x	+x	+x

Substituting the above values for equilibrium concentrations into the mass action expression and assuming that x << 0.15 gives:

$$K_a = \frac{(x)(x)}{0.15} = 6.8 \times 10^{-4}$$

$x^2 = 1.0 \times 10^{-4}$, $x = 1.0 \times 10^{-2}$ M = [H$^+$]

pH = −log[H$^+$] = −log(0.010) = 2.00

% ionization = 0.010/0.15 × 100 = 6.7 %

Since the % ionization is > 5%, we can not make the simplifying assumption. Consequently, we need to solve for x using a quadratic equation. The mass action expression is:

$$K_a = \frac{(x)(x)}{0.15 - x} = 3.0 \times 10^{-9}$$

We may rearrange this expression to obtain the following quadratic equation:

$$x^2 + 6.8 \times 10^{-4}x - 1.0 \times 10^{-4} = 0$$

where a = 1, b = 6.8 × 10^{-4}, and c = 1.0 × 10^{-4}. If we substitute these values into the quadratic equation (see Example 16.16) and take the positive root we get x = 0.0097 M.

[H$^+$] = x = 0.0097 M.

pH = −log[H$^+$] = −log(0.0097) = 2.01

% ionization = 0.0097/0.15 × 100 = 6.5 %

In this problem, the change in pH between assuming x to be small and not assuming x to be small is very slight.

16.64 $CN^- + H_2O \rightleftharpoons HCN + OH^-$

$$K_b = \frac{[HCN][OH^-]}{[CN^-]} = \frac{K_w}{K_a} = \frac{1.0 \times 10^{-14}}{3.0 \times 10^{-8}} = 1.6 \times 10^{-5}$$

	[CN$^-$]	[HCN]	[OH$^-$]
I	0.0050	—	—
C	−x	+x	+x
E	0.0050 −x	+x	+x

Assume that x << 0.0050

$$K_b = \frac{(x)(x)}{0.0050} = 1.6 \times 10^{-5} \qquad x = 2.8 \times 10^{-4} = [OH^-]$$

Wait!!!! 2.8×10^{-4} is not << 0.0050

So, use the method of successive approximations.

$$K_b = \frac{(x)(x)}{0.0050 - 0.00028} = 1.6 \times 10^{-5} \qquad x = 2.7 \times 10^{-4}$$

$$Kb = \frac{(x)(x)}{0.0050 - 0.00027} = 1.6 \times 10^{-5} \qquad x = 2.7 \times 10^{-4}$$

$$x = 2.7 \times 10^{-4} = [OH^-]$$

$$pOH = 3.56$$

$$pH = 10.44$$

16.66 $K_a = 10^{-pKa} = 10^{-4.92} = 1.2 \times 10^{-5}$

H–Paba $\rightleftharpoons$ H$^+$ + Paba$^-$

$$K_a = \frac{[H^+][Paba^-]}{[H-Paba]} = 1.2 \times 10^{-5}$$

	[H–Paba]	[H$^+$]	[Paba$^-$]
I	0.030	–	–
C	–x	+x	+x
E	0.030–x	+x	+x

Substituting the above values for equilibrium concentrations into the mass action expression and assuming that x << 0.030 gives:

$$K_a = \frac{[x][x]}{[0.030]} = 1.2 \times 10^{-5}$$

$$x^2 = 3.6 \times 10^{-7}, \quad x = 6.0 \times 10^{-4} \text{ M} = [H^+]$$

$$pH = -\log[H^+] = -\log(6.0 \times 10^{-4}) = 3.22$$

16.68   HC$_2$H$_3$O$_2$ $\rightleftharpoons$ H$^+$ + C$_2$H$_3$O$_2^-$

$$K_a = \frac{[H^+][C_2H_3O_2^-]}{[HC_2H_3O_2]} = 1.8 \times 10^{-5}$$

	[HC$_2$H$_3$O$_2$]	[H$^+$]	[C$_2$H$_3$O$_2^-$]
I	0.15	–	0.25
C	–x	+x	+x
E	0.15–x	+x	0.25+x

Substituting these values into the mass action expression gives:

$$K_a = \frac{(x)(0.25 + x)}{0.15 - x} = 1.8 \times 10^{-5}$$

Assume that x << 0.15 M and x << 0.25 M, then;

$$x \times \left(\frac{0.25}{0.15}\right) \approx 1.8 \times 10^{-5}$$

$$x \approx \left(\frac{0.15}{0.25}\right) \times 1.8 \times 10^{-5}$$

$$x \approx 1.1 \times 10^{-5} \, M = [H^+]$$

pH = −log [H$^+$] = 4.97

16.70 The equilibrium we will consider in this problem is: $NH_3 + H_2O \rightleftharpoons NH_4^+ + OH^-$

$$K_b = \frac{[NH_4^+][OH^-]}{[NH_3]} = 1.8 \times 10^{-5}$$

$$= \frac{(0.45)[OH^-]}{0.25} = 1.8 \times 10^{-5}$$

[OH$^-$] = 1.0 X 10^{-5} M
pOH = −log[OH$^-$] = −log(1.0 X 10^{-5}) = 5.00
pH = 14.00 − pOH = 14.00 − 5.00 = 9.00

16.72 The initial pH of the buffer is 4.97 as determined in Exercise 16.66. The added acid, 0.050 mol, will react with the acetate ion present in the buffer solution. Assume the added acid reacts completely. For each mole of acid added, one mole of $C_2H_3O_2^-$ is converted to $HC_2H_3O_2$. Since 0.050 mol of acid is added;

$$[HC_2H_3O_2]_{final} = (0.15 + 0.050) \, M = 0.20 \, M$$
$$[C_2H_3O_2^-]_{final} = (0.25 - 0.050) \, M = 0.20 \, M$$

Now, substitute these values into the mass action expression to calculate the final [H$^+$] in solution;

$$\frac{[H^+](0.20)}{(0.20)} = 1.8 \times 10^{-5}$$

[H$^+$] = 1.8 X 10^{-5} mol L^{-1} and the pH = 4.74

The pH of the solution changes by 4.74 − 4.93 = −0.23 pH units upon addition of the acid.

16.74 The initial pH is 9.00 as calculated in Exercise 16.70. For every mole of H$^+$ added, one mol of NH_3 will be changed to one mol of NH_4^+. Since we added 0.020 mol H$^+$;

$$[NH_4^+]_{final} = 0.45 \, M + 0.020 \, M = 0.47 \, M$$
$$[NH_3]_{final} = 0.25 \, M - 0.020 \, M = 0.23 \, M$$

Using these new concentrations, we can calculate a new pH:

$$K_b = \frac{[NH_4^+][OH^-]}{[NH_3]} = \frac{(0.47)[OH^-]}{0.23} = 1.8 \times 10^{-5}$$

[OH$^-$] = 8.8 X 10^{-6} M, the pOH = 5.06 and the pH = 8.94.

As expected, when an acid is added, the pH decreases. In this problem, the pH decreases by 0.06 pH units from 9.00 to 8.94.

16.76 $pOH = pK_b + \log\dfrac{[\text{cation}]}{[\text{base}]}$

Now pOH = 14.00 − pH = 14.00 − 9.25 = 4.75
4.75 = 4.74 + log([NH$_4^+$]/[NH$_3$])
log([NH$_4^+$]/[NH$_3$]) = 0.01
Taking the antilog of both sides of this equation gives:
[NH$_4^+$]/[NH$_3$] = $10^{0.01}$ = 1

16.78 $pH = pK_a + \log\dfrac{[\text{anion}]}{[\text{acid}]} = pK_a + \log\dfrac{[A^-]}{[HA]}$

5.00 = 4.74 + log([NaC$_2$H$_3$O$_2$]/[HC$_2$H$_3$O$_2$])
 [NaC$_2$H$_3$O$_2$]/[HC$_2$H$_3$O$_2$] = 1.8
 [NaC$_2$H$_3$O$_2$] = 1.8 [HC$_2$H$_3$O$_2$] = 1.8 0.15 = 0.27 M
Thus to the 1 L of acetic acid solution we add: 0.27 mol NaC$_2$H$_3$O$_2$ 82.0 g/mol = 22 g NaCHO$_2$.

16.80 $pOH = pK_b + \log\dfrac{[\text{cation}]}{[\text{base}]}$

Now pOH = 14.00 − pH = 14.00 − 10.00 = 4.00
4.00 = 4.74 + log([NH$_4^+$]/[NH$_3$])
log([NH$_4^+$]/[NH$_3$]) = −0.74
Taking the antilog of both sides of this equation gives:
[NH$_4^+$]/[NH$_3$] = $10^{-0.74}$ = 0.18
Thus [NH$_4^+$] = 0.18 [NH$_3$] = 0.18 0.20 M = 3.6 X 10^{-2} M = [NH$_4$Cl]

Finally, 0.500 L of buffer solution would require:
0.500 L 3.6 X 10^{-2} mol/L 53.5 g/mol = 0.96 g NH$_4$Cl

16.82 The equilibrium is; $HC_2H_3O_2 \rightleftharpoons H^+ + C_2H_3O_2^-$

$K_a = \dfrac{[H^+][C_2H_3O_2^-]}{[HC_2H_3O_2]} = 1.8 \times 10^{-5}$

The initial pH is; $\dfrac{[H^+](0.110)}{0.100} = 1.8 \times 10^{-5}$, [H$^+$] = 1.64 X 10^{-5} M and pH = 4.786. In this calculation we are able to use either the molar concentration or the number of moles since the volume is constant in this portion of the problem.

In order to calculate the change in pH, we need to determine the concentrations of HC$_2$H$_3$O$_2$ and C$_2$H$_3$O$_2^-$ after the complete reaction of the added acid. One mole of C$_2$H$_3$O$_2^-$ will be consumed for every mole of acid added and one mole of HC$_2$H$_3$O$_2$ will be produced. The number of moles of acid added is;

$\# \text{ mol H}^+ = (25.00 \text{ mL HCl})\left(\dfrac{0.100 \text{ mol HCl}}{1000 \text{ mL HCl}}\right)\left(\dfrac{1 \text{ mol H}^+}{1 \text{ mol HCl}}\right)$

$= 2.50 \times 10^{-3} \text{ mol H}^+$

The new concentration of HC$_2$H$_3$O$_2$ and C$_2$H$_3$O$_2^-$ are;

$[HC_2H_3O_2]_{\text{final}} = \dfrac{(0.100 \text{ mol} - 0.00250 \text{ mol})}{0.525 \text{ L}} = 0.195 \text{ M}$

$[C_2H_3O_2^-]_{\text{final}} = \dfrac{(0.110 \text{ mol} + 0.00250 \text{ mol})}{0.525 \text{ L}} = 0.205 \text{ M}$

Note: The new volume has been used in these calculations.

$$K_a = \frac{[H^+][C_2H_3O_2^-]}{[HC_2H_3O_2]} = \frac{[H^+](0.205)}{(0.195)} = 1.8 \times 10^{-5}$$

[H'] = 1.71 X 10⁻⁵ and pH = 4.766.

Notice that the change in pH is very small in spite of adding a strong acid. If the same amount of HCl were added to water, a completely different effect would be observed.

Since HCl is a strong acid, the [H'] in a water solution will be the result of the strong acid dissociation. We do, of course, need to account for the dilution. Using the dilution equation, $M_1V_1 = M_2V_2$, we determine the [H'] = 0.0167 mol L⁻¹ and the pH = 1.78. The change in pH in this case is 7.00 − 1.78 = 5.22 pH units. A significantly larger change!!

16.84 $H_2C_6H_6O_6 + H_2O \rightleftharpoons H_3O^+ + HC_6H_6O_6^-$ $K_{a1} = 6.7 \times 10^{-5}$
$HC_6H_6O_6^- + H_2O \rightleftharpoons H_3O^+ + C_6H_6O_6^{2-}$ $K_{a2} = 2.7 \times 10^{-12}$

	[H₂C₆H₆O₆]	[H₃O⁺]	[HC₆H₆O₆⁻]
I	0.15	–	–
C	−x	+x	+x
E	0.15−x	+x	+x

Assume x << 0.15

$$K_a = \frac{[x][x]}{[0.15]} = 6.7 \times 10^{-5} \qquad x = 3.2 \times 10^{-3}$$

$[H_2C_6H_6O_6] \cong 0.15\,M$

$[H_3O^+] = [HC_6H_6O_6^-] = 3.2 \times 10^{-3}\,M$

$[C_6H_6O_6^{2-}] = 2.7 \times 10^{-12}\,M$

pH = 2.50

pOH = 11.5

$[OH^-] = 3.2 \times 10^{-12}\,M$

16.86 $H_3PO_4 \rightleftharpoons H_2PO_4^- + H^+$ $K_{a1} = \frac{[H_2PO_4^-][H^+]}{[H_3PO_4]} = 7.1 \times 10^{-3}$

$H_2PO_4^- \rightleftharpoons HPO_4^{2-} + H^+$ $K_{a2} = \frac{[HPO_4^{2-}][H^+]}{[H_2PO_4^-]} = 6.3 \times 10^{-8}$

$HPO_4^{2-} \rightleftharpoons PO_4^{3-} + H^+$ $K_{a3} = \frac{[PO_4^{3-}][H^+]}{[HPO_4^{2-}]} = 4.5 \times 10^{-13}$

The following assumptions are made:
$[H^+]_{total} \approx [H^+]_{first\,step}$
$[H_2PO_4^-]_{total} \approx [H_2PO_4^-]_{first\,step}$
$[HPO_4^{2-}]_{total} \approx [HPO_4^{2-}]_{second\,step}$

The first dissociation:

	$[H_3PO_4]$	$[H_2PO_4^-]$	$[H^+]$
I	3.0	–	–
C	–x	+x	+x
E	3.0–x	+x	+x

$$K_{a1} = \frac{\left[H_2PO_4^-\right]\left[H^+\right]}{\left[H_3PO_4\right]} = \frac{(x)(x)}{(3.0-x)} = 7.1 \times 10^{-3}$$

Solve for x by successive approximations or by the quadratic equation:
$$x = 0.14 \text{ M} = [H^+] = [H_2PO_4^-]$$
$$[H_3PO_4] = 3.0 - 0.14 = 2.9 \text{ M}$$
$$pH = -\log(0.14) = 0.85$$

The second dissociation:

	$[H_2PO_4^-]$	$[HPO_4^{2-}]$	$[H^+]$
I	0.14	–	0.14
C	–x	+x	+x
E	0.14–x	+x	0.14+x

Assume that x << 0.14, therefore $0.14 - x \approx 0.14$ and $0.14 + x \approx 0.14$

$$K_{a2} = \frac{(x)(0.14)}{(0.14)} = 6.3 \times 10^{-8}$$
$$x = 6.3 \times 10^{-8} = [HPO_4^{2-}]$$

The third dissociation:

	$[HPO_4^{2-}]$	$[PO_4^{3-}]$	$[H^+]$
I	6.3×10^{-8}	–	0.14
C	–x	+x	+x
E	6.3×10^{-8}–x	+x	0.14+x

Assume that $x \ll 6.3 \times 10^{-8}$, therefore $6.3 \times 10^{-8} - x \approx 6.3 \times 10^{-8}$ and $0.14 + x \approx 0.14$

$$K_{a3} = \frac{(x)(0.14)}{(6.3 \times 10^{-8})} = 4.5 \times 10^{-13}$$

Solving for x we get, $x = 2.0 \times 10^{-19} = [PO_4^{3-}]$

16.88   The equilibrium is:

$$H_2Tar \;\rightleftharpoons\; HTar^- + H^+ \qquad\qquad K_{a_1} = \frac{\left[HTar^-\right]\left[H^+\right]}{\left[H_2Tar\right]} = 9.2 \times 10^{-4}$$

$$HTar^- \;\rightleftharpoons\; Tar^{2-} + H^+ \qquad\qquad K_{a_2} = \frac{\left[Tar^{2-}\right]\left[H^+\right]}{\left[HTar^-\right]} = 4.3 \times 10^{-5}$$

The $[H^+]$ and $[HTar^-]$ are determined by the first dissociation equation.

	[H₂Tar]	[HTar⁻]	[H⁺]
I	0.50	–	–
C	–x	+x	+x
E	0.50–x	+x	+x

$$K_{a_1} = \frac{\left[HTar^-\right]\left[H^+\right]}{\left[H_2Tar\right]} = \frac{(x)(x)}{(0.50-x)} = 9.2 \times 10^{-4}$$

Solving this equation requires a quadratic equation or the method of successive approximations:

$x = 0.021\ M = [H^+] = [HTar^-]$
$[H_2Tar] = 0.50 - 0.021 = 0.48\ M$
$[Tar^{2-}] = K_{a2} = 4.3 \times 10^{-5}\ M$

16.90 $H_3PO_3 \rightleftharpoons H_2PO_3^- + H^+$

$$K_{a_1} = \frac{\left[H_2PO_3^-\right]\left[H^+\right]}{\left[H_3PO_3\right]} = 1.0 \times 10^{-2}$$

$H_2PO_3^- \rightleftharpoons HPO_3^{2-} + H^+$

$$K_{a_2} = \frac{\left[HPO_3^{2-}\right]\left[H^+\right]}{\left[H_2PO_3^-\right]} = 2.6 \times 10^{-7}$$

To simplify the calculation, assume that the second dissociation does not contribute a significant amount of H⁺ to the final solution. Solving the equilibrium problem for the first dissociation gives:

	[H₃PO₃]	[H₂PO₃⁻]	[H⁺]
I	1.0	–	–
C	–x	+x	+x
E	1.0–x	+x	+x

$$K_{a_1} = \frac{\left[H_2PO_3^-\right]\left[H^+\right]}{\left[H_3PO_3\right]} = \frac{(x)(x)}{(1.0-x)} = 1.0 \times 10^{-2}$$

Because K_{a1} is so large, a quadratic equation must be solved. On doing so we learn that

$x = 0.095\ M = [H^+] = [H_2PO_3^-]$.
$pH = -\log[H^+] = -\log(0.095) = 1.022$

The $[HPO_3^{2-}]$ may be determined from the second ionization constant.

	[H₂PO₃⁻]	[HPO₃²⁻]	[H⁺]
I	0.095	–	0.095
C	–x	+x	+x
E	0.095–x	+x	0.095+x

$$K_{a_2} = \frac{\left[HPO_3^{2-}\right]\left[H^+\right]}{\left[H_2PO_3^-\right]} = \frac{(x)(0.95+x)}{(0.095-x)} = 2.6 \times 10^{-7}$$

If we assume that x is small then $0.095 \pm x \approx 0.095$. Then, $x = [HPO_3^{2-}] = 2.6 \times 10^{-7}$ M.

16.92   The hydrolysis equation is:

$$SO_3^{2-} + H_2O \rightleftharpoons HSO_3^- + OH^- \qquad\qquad K_b = \frac{\left[HSO_3^-\right]\left[OH^-\right]}{\left[SO_3^{2-}\right]}$$

In order to obtain K_b we will use the relationship $K_w = K_a \times K_b$

$$K_b = \frac{K_w}{K_a} = \frac{1.0 \times 10^{-14}}{6.6 \times 10^{-8}} = 1.5 \times 10^{-7}$$

	$[SO_3^{2-}]$	$[HSO_3^-]$	$[OH^-]$
I	0.12	–	–
C	–x	+x	+x
E	0.12–x	+x	+x

Since K_b is so small, assume that $x \ll 0.12$ and we determine $x = 1.3 \times 10^{-4}$ M = $[OH^-]$
$pOH = -\log(1.3 \times 10^{-4}) = 3.87$, $pH = 14.00 - pOH = 10.13$

16.94   Only the first hydrolysis needs to be examined. The difficulty is that we are solving for the initial concentration in this problem.

$$CO_3^{2-} + H_2O \rightleftharpoons HCO_3^- + OH^- \qquad\qquad K_b = \frac{\left[HCO_3^-\right]\left[OH^-\right]}{\left[CO_3^{2-}\right]}$$

In order to obtain K_b we will use the relationship $K_w = K_a \times K_b$

$$K_b = \frac{K_w}{K_a} = \frac{1.0 \times 10^{-14}}{4.7 \times 10^{-11}} = 2.1 \times 10^{-4}$$

	$[CO_3^{2-}]$	$[HCO_3^-]$	$[OH^-]$
I	Z	–	–
C	–x	+x	+x
E	Z–x	+x	+x

In this problem, we know that the pH at equilibrium is 11.62. Using this fact we calculate a $pOH = 2.38$ and a $[OH^-] = 10^{-pOH} = 4.2 \times 10^{-3}$ M = x. Substitute this into the equilibrium expression and solve for **Z**, our initial concentration of carbonate ion.

$$K_b = 2.1 \times 10^{-4} = \frac{\left[HCO_3^-\right]\left[OH^-\right]}{\left[CO_3^{2-}\right]} = \frac{(x)(x)}{(Z-x)} = \frac{\left(4.2 \times 10^{-3}\right)^2}{Z - 4.2 \times 10^{-3}}$$

Solving this equation we determine that $Z = 8.8 \times 10^{-2}$ M = $[CO_3^{2-}]$. (Note: We could have assumed that the change upon dissociation is small in this example, i.e., x << Z. It is unnecessary to do this for this problem. If you had made this assumption, a value of $Z = 8.4 \times 10^{-2}$ M would be obtained.)

The problem asks for the number of grams of $Na_2CO_3 \cdot 10H_2O$ so convert the concentration to a number of grams: 25.2 grams.

16.96 $C_6H_5O_7^{3-} + H_2O \rightleftharpoons HC_6H_5O_7^{2-} + OH^-$

$$K_b = \frac{\left[HC_6H_5O_7^{2-}\right]\left[OH^-\right]}{\left[C_6H_5O_7^{3-}\right]} = \frac{K_w}{K_a} = \frac{1.0 \times 10^{-14}}{4.0 \times 10^{-7}} = 2.5 \times 10^{-8}$$

	$[C_6H_5O_7^{3-}]$	$[HC_6H_5O_7^{2-}]$	$[OH^-]$
I	0.10	—	—
C	−x	+x	+x
E	0.10−x	+x	+x

Assume x << 0.10

$$K_b = \frac{(x)(x)}{0.10} = 2.5 \times 10^{-8} \qquad x = 5.0 \times 10^{-5} = [OH^-]$$

$$pOH = 4.30 \qquad pH = 9.70$$

16.98 $PO_4^{3-} + H_2O \rightleftharpoons HPO_4^{2-} + OH^-$ $\qquad K_{b1} = \dfrac{K_w}{K_{a3}} = 2.2 \times 10^{-2}$

$HPO_4^{2-} + H_2O \rightleftharpoons H_2PO_4^- + OH^-$ $\qquad K_{b2} = \dfrac{K_w}{K_{a2}} = 1.6 \times 10^{-7}$

$H_2PO_4^- + H_2O \rightleftharpoons H_3PO_4 + OH^-$ $\qquad K_{b3} = \dfrac{K_w}{K_{a1}} = 1.4 \times 10^{-12}$

By analogy with polyprotic acids, we know that
$[H_2PO_4^-] = 1.6 \times 10^{-7}$
We need to solve the first equilibrium expression to determine $[HPO_4^{2-}]$.

$$K_{b1} = \frac{\left[HPO_4^{2-}\right]\left[OH^-\right]}{\left[PO_4^{3-}\right]} = 2.2 \times 10^{-2}$$

	$[PO_4^{3-}]$	$[HPO_4^{2-}]$	$[OH^-]$
I	0.50	—	—
C	−x	+x	+x
E	0.50−x	+x	+x

Assume x << 0.50

$$K_{b1} = \frac{(x)(x)}{0.50 - x} = 2.2 \times 10^{-2}$$ Use the quadratic equation to solve for x since K_b is so large

$$x = 9.4 \times 10^{-2} \ M = [OH^-] = [HPO_4^{2-}]$$

$$pOH = 1.027$$

$$pH = 12.97$$

$[PO_4^{3-}] = 0.50 - 9.4 \times 10^{-2} = 0.41 \ M$

Solve the third equilibrium expression to determine $[H_3PO_4]$:

$$K_{b3} = \frac{[H_3PO_4][OH^-]}{[H_2PO_4^-]} = 1.4 \times 10^{-12}$$

Substitute the calculated values of $[H_2PO_4^-]$ and $[OH^-]$ and solve for x :

$$K_{b3} = \frac{(9.4 \times 10^{-2})x}{(1.6 \times 10^{-7})} = 1.4 \times 10^{-12}$$

$$x = [H_3PO_4] = 2.4 \times 10^{-18}$$

16.100   Since HCO_2H and NaOH react in a 1:1 ratio:

$$HCO_2H + NaOH \rightarrow NaHCO_2 + H_2O$$

we can use the equation $V_a \ M_a = V_b \ M_b$ to determine the volume of NaOH that is required to reach the equivalence point, i.e. the point at which the number of moles of NaOH is equal to the number of moles of HCO_2H:

$V_{NaOH} = 50 \ mL \ 0.10/0.10 = 50 \ mL$

Thus the final volume at the equivalence point will be 50 + 50 = 100 mL.

The concentration of $NaHCO_2$ would then be:

0.10 mol/L 0.050 L = 5.0×10^{-3} mol HCO_2H = 5.0×10^{-3} mol $NaHCO_2$

5.0×10^{-3} mol/0.100 L = 5.0×10^{-2} M $NaHCO_2$

The hydrolysis of this salt at the equivalence point proceeds according to the following equilibrium: $HCO_2^- + H_2O \rightleftharpoons HCO_2H + OH^-$

$$K_b = \frac{[HCO_2H][OH^-]}{[HCO_2^-]} = 5.6 \times 10^{-11}$$

	$[HCO_2^-]$	$[HCO_2H]$	$[OH^-]$
I	0.050	–	–
C	–x	+x	+x
E	0.050–x	+x	+x

Substituting the above values for equilibrium concentrations into the mass action expression and assuming x << 0.050 gives:

$$K_b = \frac{(x)(x)}{0.050} = 5.6 \times 10^{-11}$$

$x^2 = 2.8 \times 10^{-12}$ $x = 1.7 \times 10^{-6} \ M = [OH^-] = [HCO_2H]$

$pOH = -\log[OH^-] = -\log(1.7 \times 10^{-6}) = 5.77$

$pH = 14.00 - pOH = 14.00 - 5.77 = 8.23$

Cresol red would be a good indicator, since it has a color change near the pH at the equivalence point.

16.102

$$\text{\# moles HC}_2\text{H}_3\text{O}_2 = (25.0 \text{ mL HC}_2\text{H}_3\text{O}_2)\left(\frac{0.180 \text{ mol HC}_2\text{H}_3\text{O}_2}{1000 \text{ mL HC}_2\text{H}_3\text{O}_2}\right)$$

$$= 4.50 \times 10^{-3} \text{ moles HC}_2\text{H}_3\text{O}_2$$

$$\text{\# moles OH}^- = (35.0 \text{ mL OH}^-)\left(\frac{0.250 \text{ mol OH}^-}{1000 \text{ mL OH}^-}\right) = 8.75 \times 10^{-3} \text{ moles OH}^-$$

$$\text{excess OH}^- = 8.75 \times 10^{-3} - 4.50 \times 10^{-3} = 4.25 \times 10^{-3} \text{ moles}$$

$$[\text{OH}^-] = \frac{4.25 \times 10^{-3} \text{ moles}}{(25.0 + 35.0 \text{ mL})\left(\frac{1 \text{ L}}{1000 \text{ mL}}\right)} = 7.08 \times 10^{-2} \text{ M}$$

pOH = 1.150

pH = 12.850

16.104 a) $HC_2H_3O_2 \rightleftharpoons H^+ + C_2H_3O_2^- \quad K_a = \frac{[H^+][C_2H_3O_2^-]}{[HC_2H_3O_2]} = 1.8 \times 10^{-5}$

	$[HC_2H_3O_2]$	$[H^+]$	$[C_2H_3O_2^-]$
I	0.1000	–	–
C	–x	+x	+x
E	0.1000–x	+x	+x

Substituting the above values for equilibrium concentrations into the mass action expression and assuming that x << 0.1000 gives:

x = [H⁺] = 1.342 X 10⁻³ M.

pH = –log [H⁺] = –log (1.342 X 10⁻³) = 2.8724.

(b) When NaOH is added, it will react with the acetic acid present decreasing the amount in solution and producing additional acetate ion. Since this a one–to–one reaction, the number of moles of acetic acid will decrease by the same amount as the number of moles of NaOH added and the number of moles of acetate ion will increase by an identical amount. We must determine the number of moles of all ions present and calculate new concentrations accounting for dilution.

$$\text{\# moles HC}_2\text{H}_3\text{O}_2 = (0.02500 \text{ L solution})\left(\frac{0.1000 \text{ moles HC}_2\text{H}_3\text{O}_2}{1 \text{ L solution}}\right)$$

$$= 2.500 \times 10^{-3} \text{ moles HC}_2\text{H}_3\text{O}_2$$

$$\text{\# moles OH}^- = (0.01000 \text{ L solution})\left(\frac{0.1000 \text{ moles OH}^-}{1 \text{ L solution}}\right)$$

$$= 1.000 \times 10^{-3} \text{ moles OH}^-$$

$$[\text{HC}_2\text{H}_3\text{O}_2] = \frac{2.500 \times 10^{-3} \text{ moles} - 1.000 \times 10^{-3} \text{ moles}}{0.02500 \text{ L} + 0.01000 \text{ L}}$$

$$= 4.286 \times 10^{-2} \text{ M HC}_2\text{H}_3\text{O}_2$$

$$\left[C_2H_3O_2^-\right] = \frac{0 \text{ moles} + 1.000 \times 10^{-3} \text{ moles}}{0.02500 \text{ L} + 0.01000 \text{ L}}$$

$$= 2.857 \times 10^{-2} \text{ M } C_2H_3O_2^-$$

$$pH = pK_a + \log\frac{\left[C_2H_3O_2^-\right]}{\left[HC_2H_3O_2\right]}$$

$$= 4.7447 + \log\frac{\left(2.857 \times 10^{-2}\right)}{\left(4.286 \times 10^{-2}\right)} = 4.5686$$

(c)   When half the acetic acid has been neutralized, there will be equal amounts of acetic acid and acetate ion present in the solution. At this point, $pH = pK_a = 4.7447$.

(d)   At the equivalence point, all of the acetic acid will have been converted to acetate ion. The concentration of the acetate ion will be half the original concentration of acetic acid since we have doubled the volume of the solution. We then need to solve the equilibrium problem that results when we have a solution that possesses a $[C_2H_3O_2^-] = 0.05000$ M.

$$C_2H_3O_2^- + H_2O \rightleftharpoons HC_2H_3O_2 + OH^-$$

$$K_b = \frac{\left[HC_2H_3O_2\right]\left[OH^-\right]}{\left[C_2H_3O_2^-\right]} = 5.6 \times 10^{-10}$$

	$[C_2H_3O_2^-]$	$[HC_2H_3O_2]$	$[OH^-]$
I	0.05000	–	–
C	–x	+x	+x
E	0.05000–x	+x	+x

Substituting the above values for equilibrium concentrations into the mass action expression and assuming that $x \ll 0.05000$ gives: $x = [OH^-] = 5.292 \times 10^{-6}$ M.
$pOH = -\log[OH^-] = -\log(5.292 \times 10^{-6}) = 5.2764$.
$pH = 14.0000 - pOH = 14.0000 - 5.2764 = 8.7236$.

Additional Exercises

16.107   When these two solutions are mixed, the HCl will react with the ammonia producing NH_4^+. Since HCl is the limiting reactant, we will have an excess of ammonia. The resulting solution will be a buffer as it consists of ammonia, a weak base, and ammonium ion, its conjugate acid.

We need to determine the amount of NH_4^+ that is produced and the amount of NH_3 that remains:

$$\# \text{ mol } H^+ = (100 \text{ mL HCl})\left(\frac{0.500 \text{ mol HCl}}{1000 \text{ mL HCl}}\right)\left(\frac{1 \text{ mol } H^+}{1 \text{ mol HCl}}\right) = 0.0500 \text{ mol } H^+$$

$\# \text{ mol } NH_3 = \# \text{ mol initially} - \text{mol } H^+ \text{ added}$

$$= (300 \text{ mL HCl})\left(\frac{0.500 \text{ mol HCl}}{1000 \text{ mL HCl}}\right)\left(\frac{1 \text{ mol } H^+}{1 \text{ mol HCl}}\right) - 0.0500 \text{ mol } H^+$$

$$= 0.100 \text{ mol } NH_3$$

$\# \text{ mol } NH_4^+ = \# \text{ mol } H^+ = 0.0500 \text{ mol } NH_4^+$

$$pOH = pK_b + \log \frac{\left[NH_4^+\right]}{\left[NH_3\right]} = 4.74 - 0.30 = 4.44$$

$$pH = 14.00 - pOH = 9.56$$

16.110 (a)   The equation that will be used is the normal Henderson–Hasselbach equation, namely:

$$pH = pK_a + \log \frac{\left[anion\right]}{\left[acid\right]} = pK_a + \log \frac{\left[A^-\right]}{\left[HA\right]}$$

where $A^- = C_2H_3O_2^-$ and $HA = HC_2H_3O_2$. We note further that the log term involves a ratio of concentrations, but that the volume remains constant in a process such as that to be analyzed here. Thus the log term may be replaced by a ratio of mole amounts, since volumes cancel:

$$pH = pK_a + \log \frac{\left(moles\ C_2H_3O_2^-\right)}{\left(moles\ HC_2H_3O_2\right)}$$

Thus we need only determine the number of moles of acid and conjugate base that remain in the buffer after the addition of a certain amount of H' or OH⁻, in order to determine the pH of the buffer mixture after that addition.

The buffer is changed in the following way by the addition of OH⁻:

$$HC_2H_3O_2 + OH^- \rightleftharpoons H_2O + C_2H_3O_2^-$$

In other words, if 0.0100 moles of OH⁻ are added to the buffer, the amount of $HC_2H_3O_2$ goes down by 0.0100 moles, whereas the amount of $C_2H_3O_2^-$ goes up by 0.0100 moles. In addition, the pH of the solution will increase upon the addition of base.

The buffer is changed in the following way by the addition of H':

$$C_2H_3O_2^- + H' \rightleftharpoons HC_2H_3O_2$$

If 0.0100 mol of H' are added, then the amount of $C_2H_3O_2^-$ goes down by 0.0100 moles and the amount of $HC_2H_3O_2$ goes up by 0.0100 moles. As before, the addition of acid will decrease the pH of the buffer system.

For the general buffer mixture that contains x mol of $C_2H_3O_2^-$ and y mol of $HC_2H_3O_2$, we can apply the Henderson–Hasselbach equation, noting the maximum amount by which we want the pH to change (namely 0.10 units):

For the case of added base:

$$5.12 + 0.10 = 4.74 + \log \frac{\left(x + 0.0100\right)moles\ C_2H_3O_2^-}{\left(x - 0.0100\right)moles\ HC_2H_3O_2}$$

Simplifying and taking the antilog of both sides of the above equation gives: [x + 0.0100]/[y − 0.0100] = 3.0, which is now designated eq. 1.

For the case of added acid:

$$5.12 - 0.10 = 4.74 + \log \frac{\left(x - 0.0100\right)moles\ C_2H_3O_2^-}{\left(x + 0.0100\right)moles\ HC_2H_3O_2}$$

Simplifying and taking the antilog of both sides of the equation gives: [x − 0.0100]/[y + 0.0100] = 1.9, which is now designated eq. 2.

The equations 1 and 2 as designated above are solved simultaneously, since they are two equations containing two unknowns:

x = initial mol $C_2H_3O_2^-$ = 0.15 mol
y = initial mol $HC_2H_3O_2$ = 6.3 X 10^{-2} mol

These values are converted to grams as follows:
6.3×10^{-2} mol 60.1 g/mol = 3.8 g $HC_2H_3O_2$
0.15 mol $NaC_2H_3O_2$ 118 g/mol = 18 g $NaC_2H_3O_2 \cdot 2H_2O$

These are the minimum amounts of acid and conjugate base that would be required in order to prepare a buffer that would change pH by only 0.10 units on addition of either 0.0100 mol of OH^- or 0.0100 mol of H^+.

(b) 6.3×10^{-2} mol/0.250 L = 0.25 M $HC_2H_3O_2$
0.15 mol $C_2H_3O_2^-$/0.250 L = 0.60 M $NaC_2H_3O_2$

(c)   0.0100 mol OH^-/0.250 L = 4.00×10^{-2} M OH^-
$pOH = -\log[OH^-] = -\log(4.00 \times 10^{-2}) = 1.398$
$pH = 14.000 - pOH = 14.000 - 1.398 = 12.602$

(d) 1.00×10^{-2} mol H^+/0.250 L = 4.00×10^{-2} M H^+
$pH = -\log[H^+] = -\log(4.0 \times 10^{-2}) = 1.398$

16.112   Since the ammonium ion is the salt of a weak base, NH_3, it is acidic. The cyanide ion is the salt of a weak acid, HCN, so it is basic. In order to determine if the solution is acidic or basic, we need to determine the relative strength of the two components. Use the relationship $K_a \times K_b = K_w$ in order to determine the dissociation constants for the cyanide ion and the ammonium ion.

$K_a(NH_4^+) = K_w \div K_b(NH_3) = 1.0 \times 10^{-14} \div 1.8 \times 10^{-5} = 5.6 \times 10^{-10}$
$K_b(CN^-) = K_w \div K_a(HCN) = 1.0 \times 10^{-14} \div 6.2 \times 10^{-9} = 1.6 \times 10^{-5}$

Since the $K_b(CN^-)$ is larger than the $K_a(NH_4^+)$ the NH_4CN solution will be basic.

16.114   In order to solve this problem we must first neutralize the HNO_3 that is present in the solution. Since HNO_3 is a monoprotic acid and NH_3 is a monobasic base, they will react in a one to one stoichiometry. For every mole of NH_3 added, one mole of HNO_3 will be neutralized and one mole of NH_4^+ will be produced. There are initially 0.0125 moles HNO_3 = (0.250 L)(0.050 mol L^{-1}) present in the solution. Consequently, we must first add 0.0125 moles of NH_3 to neutralize all of the HNO_3.

Now the question may be rephrased to ask, how much ammonia should be added to a solution that is 0.050 M in NH_4^+ so that the final pH is 9.26?

	$[NH_3]$	$[NH_4^+]$	$[OH^-]$
I	Z	0.050	–
C	Z–x	0.050+x	+x
E	Z–x	0.050+x	+x

$$K_b = \frac{(0.050+x)(x)}{(Z-x)} = 1.8 \times 10^{-5}$$

The problem states that the equilibrium pH = 9.26 and, therefore, the pOH = 4.74 which implies that $[OH^-]$ = 1.8×10^{-5} M = x. Substituting this into the equilibrium expression and assuming that Z >> x we get

$$\frac{(0.050)(1.8 \times 10^{-5})}{Z} = 1.8 \times 10^{-5}$$

and Z = 0.050 M. Since the volume of the solution is 250 mL, this concentration corresponds to 0.0125 moles NH_3 = (0.250 L)(0.050 mol L^{-1}).

The total amount of NH_3 that must be added to the original solution is thus
0.0125 moles + 0.0125 moles = 0.0250 mols. The first addition neutralizes the HNO_3 and the second amount adjusts the pH.

The volume of NH_3 that must be added may be calculated using the ideal gas law;

$$V = \frac{nRT}{P} = \frac{(0.0250\,\text{mol})\left(0.0821\frac{\text{L atm}}{\text{mol K}}\right)(298\,\text{K})}{(740\,\text{torr})\left(\frac{1\,\text{atm}}{760\,\text{torr}}\right)} = 0.63\,\text{L} = 630\,\text{mL}$$

16.116 Ascorbic acid has the formula $H_2C_6H_6O_6$ and is a diprotic weak acid. We only need look at the first dissociation to determine the pH.

$$H_2C_6H_6O_6 \quad\rightleftharpoons\quad HC_6H_6O_6^- + H^+ \qquad\qquad K_{a_1} = \frac{\left[HC_6H_6O_6^-\right]\left[H^+\right]}{\left[H_2C_6H_6O_6\right]} = 6.8 \times 10^{-5}$$

$$\left[H_2C_6H_6O_6\right] = \left(\frac{6.0\,\text{g}\,H_2C_6H_6O_6}{250\,\text{mL}}\right)\left(\frac{1\,\text{mole}\,H_2C_6H_6O_6}{176\,\text{g}\,H_2C_6H_6O_6}\right)\left(\frac{1000\,\text{mL}}{1\,\text{L}}\right) = 0.14\,\text{M}$$

	$[H_2C_6H_6O_6]$	$[HC_6H_6O_6^-]$	$[H^+]$
I	0.14	−	−
C	−x	+x	+x
E	0.14−x	+x	+x

$$K_{a_1} = \frac{\left[HC_6H_6O_6^-\right]\left[H^+\right]}{\left[H_2C_6H_6O_6\right]} = \frac{(x)(x)}{(0.14 - x)} = 6.8 \times 10^{-5}$$

Assuming that x << 0.14 and solving we find;

x = $[H^+]$ = 3.1 X 10^{-3} M   pH = − log$[H^+]$ = − log (3.1 X 10^{-3}) = 2.51

16.118 The first dissociation gives the concentrations of H_3PO_4, H^+ and $H_2PO_4^-$:

$$H_3PO_4(aq) \rightleftharpoons H^+(aq) + H_2PO_4^-(aq)$$
$$K_{a_1} = \frac{\left[H^+\right]\left[H_2PO_4^-\right]}{\left[H_3PO_4\right]} = 7.1 \times 10^{-3}$$

	$[H_3PO_4]$	$[H_2PO_4^-]$	$[H^+]$
I	0.30	−	−
C	−x	+x	+x
E	0.30−x	+x	+x

$$K_{a_1} = \frac{\left[H^+\right]\left[H_2PO_4^-\right]}{\left[H_3PO_4\right]} = \frac{(x)(x)}{(0.30 - x)} = 7.1 \times 10^{-3}$$

Assuming that x << 0.30 and solving we find;

x = 4.6 X 10^{-2} M

A second iteration gives: x = 4.2 X 10^{-2} M
A third iteration gives: x = 4.3 X 10^{-2} M

A fourth iteration gives: $x = 4.3 \times 10^{-2}$ M

$[H_3PO_4] = 0.30 - x = 0.26$ M.
$[H^+] = [H_2PO_4^-] = x = 0.043$ M

The second dissociation gives the concentration of HPO_4^{2-}. As was shown in Example 17.1, $[HPO_4^{2-}] = K_2 = 6.3 \times 10^{-8}$ M.

The third dissociation gives the $[PO_4^{3-}]$

$$HPO_4^{2-}(aq) \rightleftharpoons H^+(aq) + PO_4^{3-}(aq)$$

$$K_{a_3} = \frac{[H^+][PO_4^{3-}]}{[HPO_4^{2-}]} = 4.5 \times 10^{-13}$$

	$[HPO_4^-]$	$[PO_4^{3-}]$	$[H^+]$
I	6.3×10^{-8}	–	0.0043
C	$-x$	$+x$	$+x$
E	$6.3 \times 10^{-8}-x$	$+x$	$0.0043+x$

$$K_{a_3} = \frac{[H^+][H_2PO_4^-]}{[H_3PO_4]} = \frac{(0.0043+x)(x)}{(6.3 \times 10^{-8} - x)} = 4.5 \times 10^{-13}$$

Assuming that $x \ll 6.3 \times 10^{-8}$ and solving we find;

$x = 6.6 \times 10^{-19}$ M $= [PO_4^{3-}]$

Practice Exercises

P17.1   (a) $K_{sp} = [Ba^{2+}][CrO_4^{2-}]$ (b) $K_{sp} = [Ag^+]^3[PO_4^{3-}]$

P17.2 $TlI \rightleftharpoons Tl^+ + I^-$

	$[Tl^+]$	$[I^-]$
I	–	–
C	+x	+x
E	+x	+x

$K_{sp} = x^2 = (1.8 \times 10^{-5})^2 = 3.2 \times 10^{-10}$

P17.3 $PbF_2(s) \rightleftharpoons Pb^{2+}(aq) + 2F^-(aq)$ $K_{sp} = [Pb^{2+}][F^-]^2$

$K_{sp} = (2.15 \times 10^{-3})(2(2.15 \times 10^{-3}))^2 = 3.98 \times 10^{-8}$

P17.4 $CoCO_3(s) \rightleftharpoons Co^{2+}(aq) + CO_3^{2-}(aq)$

	$[Co^{2+}]$	$[CO_3^{2-}]$
I	–	0.10
C	$+1.0 \times 10^{-9}$	$+ 1.0 \times 10^{-9}$
E	$+1.0 \times 10^{-9}$	$0.10 + 1.0 \times 10^{-9}$

Substituting the above values for equilibrium concentrations into the expression for K_{sp} gives:

$K_{sp} = [Co^{2+}][CO_3^{2-}] = (1.0 \times 10^{-9})(0.10 + 1.0 \times 10^{-9}) = 1.0 \times 10^{-10}$

P17.5 $PbF_2(s) \rightleftharpoons Pb^{2+}(aq) + 2F^-(aq)$

	$[Pb^{2+}]$	$[F^-]$
I	0.10	–
C	$+ 3.1 \times 10^{-4}$	$+ 2(3.1 \times 10^{-4})$
E	$0.10 + 3.1 \times 10^{-4}$	$+ 6.2 \times 10^{-4}$

Substituting the above values for equilibrium concentrations into the expression for K_{sp} gives:

$K_{sp} = [Pb^{2+}][F^-]^2 = [0.10 + 3.1 \times 10^{-4}][6.2 \times 10^{-4}]^2$
Now $(0.10 + 3.1 \times 10^{-4})$ is also ≈ 0.10:
Hence, $K_{sp} = (0.10)(6.2 \times 10^{-4})^2 = 3.9 \times 10^{-8}$

P17.6   (a) $AgBr(s) \rightleftharpoons Ag^+(aq) + Br^-(aq)$ $K_{sp} = [Ag^+][Br^-] = 5.0 \times 10^{-13}$

	$[Ag^+]$	$[Br^-]$
I	–	–
C	+ x	+ x
E	+ x	+ x

Substituting the above values for equilibrium concentrations into the expression for K_{sp} gives:

$K_{sp} = 5.0 \times 10^{-13} = [Ag^+][Br^-] = (x)(x)$

$x = \sqrt{5.0 \times 10^{-13}} = 7.1 \times 10^{-7}$

Thus the solubility is 7.1×10^{-7} M AgBr.

(b) $Ag_2CO_3(s) \rightleftharpoons 2Ag^+(aq) + CO_3^{2-}(aq)$ $K_{sp} = [Ag^+]^2[CO_3^{2-}] = 8.1 \times 10^{-12}$

	$[Ag^+]$	$[CO_3^{2-}]$
I	–	–
C	+ 2x	+ x
E	+ 2x	+ x

Substituting the above values for equilibrium concentrations into the expression for K_{sp} gives:

$K_{sp} = 8.1 \times 10^{-12} = [Ag^+]^2[CO_3^{2-}] = (2x)^2(x)$ and $4x^3 = 8.1 \times 10^{-12}$

$x = \sqrt[3]{(8.1 \times 10^{-12})/4} = 1.3 \times 10^{-4}$

Thus the molar solubility of Ag_2CO_3 is 1.3×10^{-4}.

P17.7 $AgI(s) \rightleftharpoons Ag^+(aq) + I^-(aq)$ $K_{sp} = [Ag^+][I^-] = 8.3 \times 10^{-17}$

	$[Ag^+]$	$[I^-]$
I	–	0.20
C	+ x	+ x
E	+ x	0.20 + x

Substituting the above values for equilibrium concentrations into the expression for K_{sp} gives:

$K_{sp} = 8.3 \times 10^{-17} = [Ag^+][I^-] = (x)(0.20 + x)$

We know that the value of K_{sp} is very small, and it suggests the simplifying assumption that $(0.20 + x) \approx 0.20$:

Hence, 8.3×10^{-17} (0.20)x, and $x = 4.2 \times 10^{-16}$. The assumption that $(0.20 + x)$ 0.20 is seen to be valid indeed.

Thus 4.2×10^{-16} mol of AgI will dissolve in 1.0 L of 0.20 M NaI solution.

P17.8 $Fe(OH)_3(s) \rightleftharpoons Fe^{3+}(aq) + 3OH^-(aq)$ $K_{sp} = [Fe^{3+}][OH^-]^3 = 1.6 \times 10^{-39}$

	$[Fe^{3+}]$	$[OH^-]$
I	–	0.050
C	+ x	+ 3x
E	+ x	0.050 + 3x

Substituting the above values for equilibrium concentrations into the expression for K_{sp} gives:

$K_{sp} = 1.6 \times 10^{-39} = [Fe^{3+}][OH^-]^3 = (x)[0.050 + 3x]^3$

We try to simplify by making the approximation that $(0.050 + 3x)$ 0.050:

$1.6 \times 10^{-39} = (x)(0.050)^3$ or $x = 1.3 \times 10^{-35}$

Clearly the assumption that $(0.050 + 3x) \approx 0.050$ is justified.

Thus 1.3×10^{-35} mol of $Fe(OH)_3$ will dissolve in 1.0 L of 0.050 M sodium hydroxide solution.

P17.9   The expression for K_{sp} is $K_{sp} = [Ca^{2+}][SO_4^{2-}] = 2.4 \times 10^{-5}$ and the ion product for this solution would be:
$[Ca^{2+}][SO_4^{2-}] = (2.5 \times 10^{-3})(3.0 \times 10^{-2}) = 7.5 \times 10^{-5}$

Since the ion product is larger than the value of K_{sp}, a precipitate is expected to form.

P17.10 The solubility product constant is $K_{sp} = [Ag^+]^2[CrO_4^{2-}] = 1.2 \times 10^{-12}$ and the ion product for this solution would be:
$[Ag^+]^2[CrO_4^{2-}] = (4.8 \times 10^{-5})^2(3.4 \times 10^{-4}) = 7.8 \times 10^{-13}$

Since the ion product is smaller than the value of K_{sp}, we do not expect a precipitate to form.

P17.11 We expect $PbSO_4(s)$ since nitrates are soluble.

Because two solutions are to be mixed together, there will be a dilution of the concentrations of the various ions, and the diluted ion concentrations must be used. In general, on dilution, the following relationship is found for the concentrations of the initial solution (M_i) and the concentration of the final solution (M_f): $M_i V_i = M_f V_f$

Thus the final or diluted concentrations are:

$$\left[Pb^{2+}\right] = \left(1.0 \times 10^{-3}\, M\right)\left(\frac{100.0\,mL}{200.0\,mL}\right) = 5.0 \times 10^{-4}\, M$$

$$\left[SO_4^{2-}\right] = \left(2.0 \times 10^{-3}\, M\right)\left(\frac{100.0\,mL}{200.0\,mL}\right) = 1.0 \times 10^{-3}\, M$$

The value of the ion product for the final (diluted) solution is:
$[Pb^{2+}][SO_4^{2-}] = (5.0 \times 10^{-4})(1.0 \times 10^{-3}) = 5.0 \times 10^{-7}$

Since this is smaller than the value of K_{sp} (6.3×10^{-7}), a precipitate of $PbSO_4$ is not expected.

P17.12 We expect a precipitate of $PbCl_2$ since nitrates are soluble.
We proceed as in Practice Exercise 13. $M_i V_i = M_f V_f$

$$\left[Pb^{2+}\right] = \left(0.10\, M\right)\left(\frac{50.0\,mL}{70.0\,mL}\right) = 0.071 M$$

$$\left[Cl^-\right] = \left(0.040\, M\right)\left(\frac{20.0\,mL}{70.0\,mL}\right) = 0.011 M$$

The value of the ion product for such a solution would be:
$[Pb^{2+}][Cl^-]^2 = (7.1 \times 10^{-2})(1.1 \times 10^{-2})^2 = 8.6 \times 10^{-6}$

Since the ion product is smaller than K_{sp}, we expect that no precipitate of $PbCl_2$ can form.

P17.13 In a saturated hydrogen sulfide solution, $[H_2S] = 0.10$ M. The appropriate equilibrium is:
$$FeS(s) + 2H^+ \rightleftharpoons Fe^{2+} + H_2S$$

The mass action expression is:

$$K_{spa} = \frac{[Fe^{2+}][H_2S]}{[H^+]^2} = 6 \times 10^2$$

The other equilibrium to consider is $HgS(s) + 2H^+ \rightleftharpoons Hg^{2+} + H_2S$

The mass action expression for this equilibrium is:

$$K_{spa} = \frac{[Hg^{2+}][H_2S]}{[H^+]^2} = 2 \times 10^{-32}$$

It is easy to see that HgS is much less soluble than FeS. We need to determine the $[H^+]$ at which FeS starts to precipitate. Rearranging the equilibrium expression we get:

$$[H^+] = \sqrt{\frac{[Fe^{2+}][H_2S]}{K_{spa}}} = \sqrt{\frac{(0.010)(0.10)}{600}} = 1.3 \times 10^{-3}\ M$$

$pH = -\log[H^+] = -\log(1.3 \times 10^{-3}) = 2.9$

Above this pH, the FeS becomes insoluble.
(Note: The value reported for the $[H^+]$ has too many sig figs. However, the value for the pH has the correct number of sig figs. This was done simply to make the values more readable.)

P17.14 Follow the exact procedure outlined in example 17.10.
$K_{sp} = [Ca^{2+}][CO_3^{2-}] = 4.5 \times 10^{-9}$
$K_{sp} = [Ni^{2+}][CO_3^{2-}] = 1.3 \times 10^{-7}$

$NiCO_3$ is more soluble and will precipitate when:
$$[CO_3^{2-}] = \frac{K_{sp}}{[Ni^{2+}]} = \frac{1.3 \times 10^{-7}}{0.10} = 1.3 \times 10^{-6}$$
$CaCO_3$ will precipitate when:
$$[CO3] = \frac{Ksp}{[Ca2+]} = \frac{4.5 \times 10^{-9}}{0.10} = 4.5 \times 10^{-8}$$

$CaCO_3$ will precipitate and $NiCO_3$ will not precipitate if $[CO_3^{2-}]$ .4.5 x 10^{-8} and $[CO_3^{2-}] < 1.3 \times 10^{-6}$. Now, using the equation highlighted in example 17.10 we get:

$$[H^+]^2 = (2.4 \times 10^{-17})\left(\frac{0.030}{[CO_3^{2-}]}\right)\quad NiCO_3\ \text{will precipitate if}$$

$$[H^+]^2 = (2.4 \times 10^{-17})\left(\frac{0.030}{1.3 \times 10^{-6}}\right) = 5.5 \times 10^{-13}$$

$[H^+] = 7.4 \times 10^{-7}\qquad pH = 6.13$

$CaCO_3$ will precipitate if :

$$[H^+]^2 = (2.4 \times 10^{-17})\left(\frac{0.030}{4.5 \times 10^{-8}}\right) = 1.6 \times 10^{-11}$$

$[H^+] = 4.0 \times 10^{-6}\qquad pH = 5.40$

So $CaCO_3$ will precipitate and $NiCO_3$ will not if the pH is maintained between pH = 5.40 and pH = 6.13

P17.15   The overall equilibrium is $AgCl(s) + 2NH_3(aq) \rightleftharpoons Ag(NH_3)_2^+(aq) + Cl^-(aq)$

$$K_c = \frac{\left[Ag(NH_3)_2^+\right]\left[Cl^-\right]}{[NH_3]^2}$$

In order to obtain a value for K_c for this reaction, we need to use the expressions for K_{sp} of AgCl(s) and the K_{form} of $Ag(NH_3)_2^+$:

$$K_{sp} = \left[Ag^+\right]\left[Cl^-\right] = 1.8 \times 10^{-10}$$

$$K_{form} = \frac{\left[Ag(NH_3)_2^+\right]}{\left[Ag^+\right]\left[NH_3\right]^2} = 1.6 \times 10^7$$

$$K_c = K_{sp} \times K_{form} = \frac{\left[Ag(NH_3)_2^+\right]\left[Cl^-\right]}{[NH_3]^2} = 2.9 \times 10^{-3}$$

Now we may use an equilibrium table for the reaction in question:

	$[NH_3]$	$[Ag(NH_3)_2^+]$	$[Cl^-]$
I	0.10	—	—
C	$-2x$	$+x$	$+x$
E	$0.10-2x$	$+x$	$+x$

Substituting these values into the mass action expression gives:

$$K_c = 2.9 \times 10^{-3} = \frac{(x)(x)}{(0.10-2x)^2}$$

Take the square root of both sides to get $0.054 = \frac{(x)}{(0.10-2x)}$

Solving for x we get, $x = 4.9 \times 10^{-3}$ M. The molar solubility of AgCl in 0.10 M NH_3 is therefore 4.9×10^{-3} M.

In order to determine the solubility in pure water, we simply look at K_{sp}

$$AgCl(s) \rightleftharpoons Ag^+(aq) + Cl^-(aq) \qquad K_{sp} = [Ag^+][Cl^-] = 1.8 \times 10^{-10}$$

At equilibrium; $[Ag^+] = [Cl^-] = 1.3 \times 10^{-5}$. Hence the the molar solubility of AgCl in 0.10 M NH_3 is about 380 times greater than in pure water.

P17.16   We will use the information gathered for the last problem. Specifically,

$$AgCl(s) + 2NH_3(aq) \rightleftharpoons Ag(NH_3)_2^+(aq) + Cl^-(aq)$$

$$K_c = \frac{\left[Ag(NH_3)_2^+\right]\left[Cl^-\right]}{[NH_3]^2} = 2.9 \times 10^{-3}$$

If we completely dissolve 0.20 mol of AgCl, the equilibrium $[Cl^-]$ and $[Ag(NH_3)_2^+]$ will be 0.20 M in a one liter container. The question asks, therefore, what amount of NH_3 must be initially present so that the equilibrium concentration of Cl^- is 0.20M?

	$[NH_3]$	$[Ag(NH_3)_2{}^+]$	$[Cl^-]$
I	Z	–	–
C	–2x	+x	+x
E	Z–2x	+x	+x

$$K_c = 2.9 \times 10^{-3} = \frac{(x)(x)}{(Z-2x)^2}$$

Take the square root of both sides to get;

$$0.054 = \frac{x}{Z-2x} = \frac{0.20}{Z-0.40}$$

We have substituted the known value of x. Solving for Z we get, Z = 4.1 M

Consequently, we would need to add 4.1 moles of NH_3 to a one liter container of 0.20 M AgCl in order to completely dissolve the AgCl.

Review Problems

17.9   (a) $CaF_2(s) \rightleftharpoons Ca^{2+} + 2F^-$ $K_{sp} = [Ca^{2+}][F^-]^2$

        (b) $Ag_2CO_3(s) \rightleftharpoons 2Ag^+ + CO_3{}^{2-}$ $K_{sp} = [Ag^+]^2[CO_3{}^{2-}]$

        (c) $PbSO_4(s) \rightleftharpoons Pb^{2+} + SO_4{}^{2-}$ $K_{sp} = [Pb^{2+}][SO_4{}^{2-}]$

        (d) $Fe(OH)_3(s) \rightleftharpoons Fe^{3+} + 3OH^-$ $K_{sp} = [Fe^{3+}][OH^-]^3$

        (e) $PbI_2(s) \rightleftharpoons Pb^{2+} + 2I^-$ $K_{sp} = [Pb^{2+}][I^-]^2$

        (f) $Cu(OH)_2(s) \rightleftharpoons Cu^{2+} + 2OH^-$ $K_{sp} = [Cu^{2+}][OH^-]^2$

17.11

$$\text{\# moles BaSO}_4 = (0.00245 \text{ g BaSO}_4)\left(\frac{1 \text{ mole BaSO}_4}{233.3906 \text{ g BaSO}_4}\right)$$

$$\text{\# moles BaSO}_4 = 1.05 \times 10^{-5} \text{ moles}$$

$$[Ba^{2+}] = [SO_4{}^{2-}] = 1.05 \times 10^{-5} \text{ M}$$

$$K_{sp} = [Ba^{2+}][SO_4{}^{2-}] = (1.05 \times 10^{-5})^2 = 1.10 \times 10^{-10}$$

17.13 $BaSO_3(s) \rightleftharpoons Ba^{2+} + SO_3{}^{2-}$ $K_{sp} = [Ba^{2+}][SO_3{}^{2-}]$
        $K_{sp} = (0.10)(8.0 \times 10^{-6}) = 8.0 \times 10^{-7}$
        In this problem, all of the Ba^{2+} comes from the $BaCl_2$.

17.15 $Ag_3PO_4(s) \rightleftharpoons 3Ag^+ + PO_4{}^{3-}$ $K_{sp} = [Ag^+]^3[PO_4{}^{3-}]$
        $K_{sp} = (3(1.8 \times 10^{-5}))^3(1.8 \times 10^{-5}) = 2.8 \times 10^{-18}$

17.17 $PbBr_2(s) \rightleftharpoons Pb^{2+} + 2Br^-$ $K_{sp} = [Pb^{2+}][Br^-]^2$

	$[Pb^{2+}]$	$[Br^-]$
I	–	–
C	+ x	+ 2x
E	+ x	+ 2x

$$K_{sp} = (x)(2x)^2 = 4x^3 = 2.1 \times 10^{-6}, \quad x = \sqrt[3]{\frac{2.1 \times 10^{-6}}{4}} = 8.1 \times 10^{-3} \text{ M}$$

17.19   For every mole of CO_3^{2-} produced, 2 moles of Ag^+ will be produced. Let $x = \left[CO_3^{2-}\right]$ at equilibrium and $\left[Ag^+\right] = 2x$ at equilibrium. $K_{sp} = \left[Ag^+\right]^2\left[CO_3^{2-}\right] = (2x)^2(x) = 4x^3$. Solving we find $x = 1.3 \times 10^{-4}$. Thus, the molar solubility of Ag_2CO_3 is 1.3×10^{-4} moles/L.

17.21   To solve this problem, determine the molar solubility for each compound.

LiF: let $x = \left[Li^+\right] = \left[F^-\right]$ $K_{sp} = \left[Li^+\right]\left[F^-\right] = x^2 = 1.7 \times 10^{-3}$
$x = 4.1 \times 10^{-2}$ moles/L = molar solubility of LiF.

BaF_2:   let $x = \left[Ba^{2+}\right]$, $\left[F^-\right] = 2x$ $K_{sp} = \left[Ba^{2+}\right]\left[F^-\right]^2 = (x)(2x)^2 = 1.7 \times 10^{-6}$
$4x^3 = 1.7 \times 10^{-6}$, and $x = 7.5 \times 10^{-3}$ M = molar solubility of BaF_2.

Because the molar solubility of LiF is greater than the molar solubility of BaF_2, LiF is more soluble.

17.23   First determine the molar solubility of the MX salt.
Let $x = \left[M^+\right] = \left[X^-\right]$, $K_{sp} = \left[M^+\right]\left[X^-\right] = (x)(x) = 3.2 \times 10^{-10}$
$x = 1.8 \times 10^{-5}$ M. This is the equilibrium concentration of the two ions.

For the MX_3 salt, let x = equilibrium concentration of M^{3+}, $\left[X^-\right] = 3x$.
$K_{sp} = \left[M^{3+}\right]\left[X^-\right]^3 = (x)(3x)^3 = 27x^4$. The value of x in this expression is the value determined in the first part of this problem.
So, $K_{sp} = (27)(1.8 \times 10^{-5})^4 = 2.8 \times 10^{-18}$

17.25 $CaSO_4(s) \rightleftharpoons Ca^{2+}(aq) + SO_4^{2-}(aq)$ $K_{sp} = \left[Ca^{2+}\right]\left[SO_4^{2-}\right]$

let $x = \left[Ca^{2+}\right] = \left[SO_4^{2-}\right]$ $K_{sp} = x^2 = 2.4 \times 10^{-5}$ and $x = 4.9 \times 10^{-3}$ M.
The molar solubility of $CaSO_4$ is 4.9×10^{-3} moles/L.

17.27   (a) $CuCl(s) \rightleftharpoons Cu^+(aq) + Cl^-(aq)$ $K_{sp} = \left[Cu^+\right]\left[Cl^-\right]$

	$[Cu^+]$	$[Cl^-]$
I	–	–
C	+x	+x
E	+x	+x

$K_{sp} = x^2 = 1.9 \times 10^{-7}$ $\therefore$ x = molar solubility = 4.4×10^{-4} M

(b) $CuCl(s) \rightleftharpoons Cu^+(aq) + Cl^-(aq)$ $K_{sp} = \left[Cu^+\right]\left[Cl^-\right]$

	$[Cu^+]$	$[Cl^-]$
I	–	0.0200
C	+x	+x
E	+x	0.0200+x

$K_{sp} = (x)(0.0200+x) = 1.9 \times 10^{-7}$ Assume that x << 0.0200
$\therefore$ x = molar solubility = 9.5×10^{-6} M

(c) $CuCl(s) \rightleftharpoons Cu^+(aq) + Cl^-(aq)$ $K_{sp} = \left[Cu^+\right]\left[Cl^-\right]$

	$[Cu^+]$	$[Cl^-]$
I	–	0.200
C	+x	+x
E	+x	0.200+x

$K_{sp} = (x)(0.200+x) = 1.9 \times 10^{-7}$ Assume that x << 0.200
$\therefore$ x = molar solubility = 9.5×10^{-7} M

(d) $CuCl(s) \rightleftharpoons Cu^+(aq) + Cl^-(aq)$ $K_{sp} = \left[Cu^+\right]\left[Cl^-\right]$

Note that the Cl^- concentration equals (2)(0.150 M) since two moles of Cl^- are produced for every mole of $CaCl_2$.

	$[Cu^+]$	$[Cl^-]$
I	–	0.300
C	+x	+x
E	+x	0.300+x

$K_{sp} = (x)(0.300+x) = 1.9 \times 10^{-7}$ Assume that x << 0.300
$\therefore$ x = molar solubility = 6.3×10^{-7} M

17.29 $Ag_2CrO_4(s) \rightleftharpoons 2Ag^+(aq) + CrO_4^{2-}(aq)$ $K_{sp} = \left[Ag^+\right]^2\left[CrO_4^{2-}\right]$

	$[Ag^+]$	$[CrO_4^{2-}]$
I	0.200	–
C	0.200+2x	+x
E	0.200+2x	+x

$K_{sp} = (0.200+2x)^2(x)$ Assume that x << 0.200
$1.2 \times 10^{-12} = (0.200)^2(x)$ $x = 3.0 \times 10^{-11}$
The molar solubility is 3.0×10^{-11} moles/L.

17.31 $CaSO_4(s) \rightleftharpoons Ca^{2+}(aq) + SO_4^{2-}(aq)$ $K_{sp} = \left[Ca^{2+}\right]\left[SO_4^{2-}\right]$
let x = $[Ca^{2+}]$, $[SO_4^{2-}]$ = 0.015 M + x K_{sp} = (x)(0.015 + x) = 2.4×10^{-5}
Solving the resulting quadratic equation gives x = 1.5×10^{-3} M.
The molar solubility of $CaSO_4$ in 0.015 M $CaCl_2$ is 1.5×10^{-3} moles/L.
(Note: If we had assumed that x << 0.015, as is usually the method we use to solve these problems, we would have determined that x = 0.0016 M. This is slightly larger than 10% of the value 0.015 so our usual assumption is not valid in this problem.)

17.33 In order for a precipitate to form, the value of the reaction quotient, Q, must be greater than the value of K_{sp}. For $PbCl_2$, $K_{sp} = 1.7 \times 10^{-5}$ (see Table).

$$Q = \left[Pb^{2+}\right]\left[Cl^-\right]^2 = (0.0150)(0.0120)^2 = 2.16 \times 10^{-6}.$$ Since $Q < K_{sp}$, no precipitate will form.

17.35 To solve this problem, determine the value for Q and apply LeChâtelier's Principle.

(a) $\left[Pb^{2+}\right] = (50.0 \text{ mL})(0.0100 \text{ moles/L})/(100.0 \text{ mL}) = 5.00 \times 10^{-3}$

$\left[Br^-\right] = (50.0 \text{ mL})(0.0100 \text{ moles/L})/(100.0 \text{ mL}) = 5.00 \times 10^{-3}$

$Q = \left[Pb^{2+}\right]\left[Br^-\right]^2 = (5.00 \times 10^{-3})(5.00 \times 10^{-3})^2 = 1.25 \times 10^{-7}$

For $PbBr_2$, $K_{sp} = 2.1 \times 10^{-6}$

Since $Q < K_{sp}$, no precipitate will form.

(b) $\left[Pb^{2+}\right] = (50.0 \text{ mL})(0.0100 \text{ moles/L})/(100.0 \text{ mL}) = 5.00 \times 10^{-3}$

$\left[Br^-\right] = (50.0 \text{ mL})(0.100 \text{ moles/L})/(100.0 \text{ mL}) = 5.00 \times 10^{-2}$

$Q = \left[Pb^{2+}\right]\left[Br^-\right]^2 = (5.00 \times 10^{-3})(5.00 \times 10^{-2})^2 = 1.25 \times 10^{-5}$

For $PbBr_2$, $K_{sp} = 2.1 \times 10^{-6}$

Since $Q > K_{sp}$, a precipitate will form.

17.37 $AgCl(s) \rightleftharpoons Ag^+ + Cl^-$ ‎ ‎ ‎ ‎ $K_{sp} = \left[Ag^+\right]\left[Cl^-\right] = 1.8 \times 10^{-10}$

$AgI(s) \rightleftharpoons Ag^+ + I^-$ ‎ ‎ ‎ ‎ $K_{sp} = \left[Ag^+\right]\left[I^-\right] = 8.3 \times 10^{-17}$

When $AgNO_3$ is added to the solution, AgI will precipitate before any AgCl does due to the lower solubility of AgI. In order to answer the question, i.e., what is the $[I^-]$ when AgCl first precipitates, we need to find the minimum concentration of Ag^+ that must be added to precipitate AgCl.

Let $x = [Ag^+]$; $K_{sp} = (x)(0.050) = 1.8 \times 10^{-10}$; $x = 3.6 \times 10^{-9}$ M

When the AgCl starts to precipitate, the solution will have a $[Ag^+]$ of 3.6×10^{-9} M. Now we ask, what is the $[I^-]$ if $[Ag^+] = 3.6 \times 10^{-9}$ M?

So, $K_{sp} = \left[Ag^+\right]\left[I^-\right] = (3.6 \times 10^{-9})(x) = 8.3 \times 10^{-17}$; $x = 2.3 \times 10^{-8}$ M $= [I^-]$

17.39 The less soluble substance is PbS. We need to determine the minimum $[H^+]$ at which CoS will precipitate.

$$K_{spa} = \frac{\left[Co^{2+}\right]\left[H_2S\right]}{\left[H^+\right]^2} = \frac{(0.010)(0.1)}{[H^+]^2} = 0.5 \text{ (from Table 17.3)}$$

$$[H^+] = \sqrt{\frac{(0.010)(0.1)}{0.5}} = 0.045$$

$pH = -\log[H^+] = 1.35$. At a pH lower than 1.35, PbS will precipitate and CoS will not. At larger values of pH, both PbS and CoS will precipitate.

17.41 (a) $Cu^{2+}(aq) + 4Cl^-(aq) \rightleftharpoons CuCl_4^{2-}(aq)$

$$K_{form} = \frac{\left[CuCl_4^{2-}\right]}{\left[Cu^{2+}\right]\left[Cl^-\right]^4}$$

(b) $\quad Ag^+(aq) + 2I^-(aq) \rightleftharpoons AgI_2^-(aq)$

$$K_{form} = \frac{\left[AgI_2^-\right]}{\left[Ag^+\right]\left[I^-\right]^2}$$

(c) $\quad Cr^{3+}(aq) + 6NH_3(aq) \rightleftharpoons Cr(NH_3)_6^{3+}(aq)$

$$K_{form} = \frac{\left[Cr(NH_3)_6^{3+}\right]}{\left[Cr^{3+}\right]\left[NH_3\right]^6}$$

17.43 (a) $\quad Co(NH_3)_6^{3+}(aq) \rightleftharpoons Co^{3+}(aq) + 6NH_3(aq)$

$$K_{inst} = \frac{\left[Co^{3+}\right]\left[NH_3\right]^6}{\left[Co(NH_3)_6^{3+}\right]}$$

(b) $\quad HgI_4^{2-}(aq) \rightleftharpoons Hg^{2+}(aq) + 4I^-(aq)$

$$K_{inst} = \frac{\left[Hg^{2+}\right]\left[I^-\right]^4}{\left[HgI_4^{2-}\right]}$$

(c) $\quad Fe(CN)_6^{4-}(aq) \rightleftharpoons Fe^{2+}(aq) + 6CN^-(aq)$

$$K_{inst} = \frac{\left[Fe^{2+}\right]\left[CN^-\right]^6}{\left[Fe(CN)_6^{4-}\right]}$$

17.45 There are two events in this net process: one is the formation of a complex ion (an equilibrium which has an appropriate value for K_{form}), and the other is the dissolving of $Fe(OH)_3$, which is governed by K_{sp} for the solid.

$Fe(OH)_3(s) \rightleftharpoons Fe^{3+}(aq) + 3OH^-(aq)$ $\qquad K_{sp} = \left[Fe^{3+}\right]\left[OH^-\right]^3 = 1.6 \times 10^{-39}$

$Fe^{3+}(aq) + 6CN^-(aq) \rightleftharpoons Fe(CN)_6^{3-}(aq)$ $\qquad K_{form} = \frac{\left[Fe(CN)_6^{3-}\right]}{\left[Fe^{3+}\right]\left[CN^-\right]^6} = 1.0 \times 10^{31}$

The net process is:

$Fe(OH)_3(s) + 6CN^-(aq) \rightleftharpoons Fe(CN)_6^{3-}(aq) + 3OH^-(aq)$

The equilibrium constant for this process should be:

$$K_c = \frac{\left[Fe(CN)_6^{3-}\right]\left[OH^-\right]^3}{\left[CN^-\right]^6}$$

The numerical value for the above K_c is equal to the product of K_{sp} for $Fe(OH)_3(s)$ and K_{form} for $Fe(CN)_6^{3-}$, as can be seen by multiplying the mass action expressions for these two equilibria: $K_c = K_{form} \times K_{sp} = 1.6 \times 10^{-8}$

Because K_{form} is so very large, we can assume that all of the dissolved iron ion is present in solution as the complex, thus: $[Fe(CN)_6^{3-}] = 0.11$ mol/1.2 L = 0.092 M. Also the reaction stoichiometry shows that each iron ion that dissolves gives 3 OH^- ions in solution, and we have: $[OH^-] = 0.092 \times 3 = 0.28$ M. We substitute these values into the K_c expression and rearrange to get:

$$[CN-] = \sqrt[6]{\frac{\left[Fe(CN)_6^{3-}\right]\left[OH^-\right]^3}{K_c}}$$

$$= \sqrt[6]{\frac{(0.092)(0.28)^3}{1.6 \times 10^{-8}}}$$

Thus we arrive at the concentration of cyanide ion that is required in order to satisfy the mass action requirements of the equilibrium: $[CN^-] = 7.1$ mol L^{-1}. Since this concentration of CN^- must be present in 1.2 L, the number of moles of cyanide that are required is: 7.1 mol $L^{-1} \times 1.2$ L = 8.5 mol CN^-.

Additionally, a certain amount of cyanide is needed to form the complex ion. The stoichiometry requires six times as much cyanide ion as iron ion. This is 0.11 moles $\times$ 6 = 0.66 mol. This brings the total required cyanide to (8.5 + 0.66) = 9.2 mol.

9.2 mol 49.0 g/mol = 450 g NaCN are required.

Additional Exercises

17.48 $FeS(s) + 2H^+(aq) \rightleftharpoons Fe^{2+}(aq) + H_2S(aq)$ $K_{spa} = \dfrac{\left[Fe^{2+}\right]\left[H_2S\right]}{\left[H^+\right]^2}$

	[H⁺]	[Fe²⁺]	[H₂S]
I	8	—	—
C	8–2x	+x	+x
E	8–2x	+x	+x

$$K_{spa} = \frac{\left[Fe^{2+}\right]\left[H_2S\right]}{\left[H^+\right]^2} = \frac{(x)(x)}{(8-2x)^2} = 600 \text{ (from Table 17.3)}$$

take the square root of both sides to get; $\dfrac{x}{(8-2x)} = 24.5$

Solving gives x = 3.92 M. FeS is very soluble in 8 M acid.

17.50   (a)   The number of moles of the two reactants are:

0.12 M $Ag^+ \times 0.050$ L = 6.0×10^{-3} moles Ag^+
0.048 M $Cl^- \times 0.050$ L = 2.4×10^{-3} moles Cl^-

The precipitation of AgCl proceeds according to the following stoichiometry:
$Ag^+ + Cl^- \rightarrow AgCl(s)$. If we assume that the product is completely insoluble, then 2.4×10^{-3} moles of AgCl will be formed because Cl^- is the limiting reagent (see above.)

$$\# \text{ g AgCl} = \left(2.4 \times 10^{-3} \text{ mol AgCl}\right)\left(\frac{143.3 \text{ g AgCl}}{1 \text{ mol AgCl}}\right) = 0.35 \text{ g AgCl}$$

(b)   The silver ion concentration may be determined by calculating the amount of excess silver added to the solution:

$[Ag^+] = (6.0 \times 10^{-3}$ moles $- 3.6 \times 10^{-2}$ moles$)/1.00$ L = 3.6×10^{-2} M

The concentrations of nitrate and sodium ions are easily calculated since they are spectators in this reaction:

$[NO_3^-] = (0.12 \text{ M})(50.0 \text{ mL})/(100.0 \text{ mL}) = 6.0 \times 10^{-2}$ M
$[Na^+] = (0.048 \text{ M})(50.0 \text{ mL})/(100.0 \text{ mL}) = 2.4 \times 10^{-2}$ M

In order to determine the chloride ion concentration, we need to solve the equilibrium expression. Specifically, we need to ask what is the chloride ion concentration in a saturated solution of AgCl that has a $[Ag^+] = 3.6 \times 10^{-2}$ M.

$$AgCl(s) \rightleftharpoons Ag^+ + Cl^- \qquad K_{sp} = 1.8 \times 10^{-10}$$

	$[Ag^+]$	$[Cl^-]$
I	0.036	–
C	+x	+x
E	0.036+x	+x

$$K_{sp} = \left[Ag^+\right]\left[Cl^-\right] = (0.036+x)(x) = 1.8 \times 10^{-10}$$
$$x = 5.0 \times 10^{-9} \text{ M if we assume that } x << 0.036$$

Therefore, $[Cl^-] = 5.0 \times 10^{-9}$ M.

(c)   The percentage of the silver that has precipitated is:

$$(2.4 \times 10^{-3} \text{ moles})/(6.0 \times 10^{-3} \text{ moles}) \times 100\% = 40\%$$

17.52   A saturated solution of $La_2(CO_3)_3$ satisfies the following equilibrium expression:

$$K_{sp} = 4.0 \times 10^{-34} = \left[La^{3+}\right]^2\left[CO_3^{2-}\right]^3$$

If $\left[La^{3+}\right] = 0.010$ M, then the carbonate concentration of a saturated solution is:

$$\left[CO_3^{2-}\right] = \sqrt[3]{\frac{K_{sp}}{\left[La^{3+}\right]^2}} = \sqrt[3]{\frac{4.0 \times 10^{-34}}{(0.010)^2}} = 1.6 \times 10^{-10}$$

We do the same calculation for $PbCO_3$:

$$K_{sp} = 7.4 \times 10^{-14} = \left[Pb^{2+}\right]\left[CO_3^{2-}\right]$$

If $\left[Pb^{2+}\right] = 0.010$ M, then the carbonate concentration of a saturated solution is:

$$\left[CO_3^{2-}\right] = \frac{K_{sp}}{\left[Pb^{2+}\right]} = \frac{7.4 \times 10^{-14}}{0.010} = 7.4 \times 10^{-12} \text{ M}$$

Therefore, at a carbonate ion concentration between 7.4×10^{-12} M and 1.6×10^{-10} M, $PbCO_3$ will precipitate, but $La_2(CO_3)_3$ will not precipitate. The upper limit for the carbonate ion concentration is therefore 1.6×10^{-10} M.

The equilibrium we need to look at now is: $H_2CO_3(aq) \rightleftharpoons 2H^+(aq) + CO_3^{2-}(aq)$

The K_a for this reaction is the product of K_{a_1} and K_{a_2} for carbonic acid. From Table 17.1 we see that $K_{a_1} = 4.5 \times 10^{-7}$ and $K_{a_2} = 4.7 \times 10^{-11}$. So the equilibrum expression and value for the reaction of interest is:

$$K_a = \frac{[H^+]^2 [CO_3{}^{2-}]}{[H_2CO_3]} = K_{a_1} \times K_{a_2} = 2.1 \times 10^{-17}$$

This equation is rearranged and the values above and the values given in the problem are substituted in order to determine the pH range over which $PbCO_3$ will selectively precipitate:

$$[H^+] = \sqrt{\frac{K_a [H_2CO_3]}{[CO_3{}^{2-}]}} = \sqrt{\frac{(2.1 \times 10^{-17})(3.3 \times 10^{-2})}{[CO_3{}^{2-}]}}$$

If we substitute $[CO_3{}^{2-}] = 7.4 \times 10^{-12}$ M we determine $[H^+] = 3.1 \times 10^{-4}$ M and the pH = 3.51. Substituting $[CO_3{}^{2-}] = 1.6 \times 10^{-10}$ M, $[H^+] = 6.6 \times 10^{-5}$ M and pH = 4.18.

Consequently, if $[H^+] = 6.6 \times 10^{-5}$ M (pH = 4.18), $La_2(CO_3)_3$ will not precipitate but $PbCO_3$ will precipitate. At pH = 3.51 and below, neither carbonate will precipitate.

17.54 There are two reactions that have to be considered here: the dissociation of $CaCO_3$ in water,
$$CaCO_3(s) \rightarrow Ca^{2+}(aq) + CO_3{}^{2-}(aq) \quad K_{sp} = [Ca^{2+}][CO_3{}^{2-}] = 4.5 \times 10^{-9}$$
and the ionization of carbonate ion in water,

$$CO_3{}^{2-}(aq) + H_2O \rightarrow HCO_3{}^-(aq) + OH^-(aq) \quad K_b = \frac{[OH^-][HCO_3{}^-]}{[CO_3{}^{2-}]} = 1.8 \times 10^{-4}$$

Assuming that all of the $CO_3{}^{2-}$ reacts with the water, the net reaction is:
$$CaCO_3(s) \rightarrow Ca^{2+}(aq) + HCO_3{}^-(aq) + OH^-(aq)$$
$$K_c = K_{sp} \times K_b = [Ca^{2+}][HCO_3{}^-][OH^-] = 8.1 \times 10^{-13}$$

We can obtain $[OH^-]$ from the pH:
$$pOH = 14 - pH = 14 - 8.50 = 5.50$$
$$[OH^-] = 10^{-pOH} = 10^{-5.50} = 3.2 \times 10^{-6} \text{ M}$$
Assume that $[Ca^{2+}] = [HCO_3{}^-] = x$
$$K_c = 8.1 \times 10^{-13} = (x)(3.2 \times 10^{-6} \text{ M})(x) = (x)^2(3.2 \times 10^{-6} \text{ M})$$
Solve for x:
$$x = 5.0 \times 10^{-4} \text{ M} = [Ca^{2+}] = [HCO_3{}^-]$$
The $[Ca^{2+}]$ is equal to the molar solubility. Thus, the molar solubility of $CaCO_3$ is 5.0×10^{-4} M.

17.57 The reaction for this problem is the formation of $Ag(NH_3)_2{}^+$:
$$Ag^+(aq) + 2NH_3(aq) \rightarrow Ag(NH_3)_2{}^+ \quad K_{form} = \frac{[Ag(NH_3)_2{}^+]}{[Ag^+][NH_3]^2} = 1.6 \times 10^7$$

We can rearrange this equation and substitute the values from Example 19.2 to determine the $[Ag^+]$:

$$[Ag^+] = \frac{[Ag(NH_3)_2{}^+]}{K_{form}[NH_3]^2} = \frac{(2.8 \times 10^{-3})}{(1.6 \times 10^7)(1)^2} = 1.8 \times 10^{-10} \text{ M}$$

Practice Exercises

P18.1   (a) $q>0$, $w>0$
      (b) $q<0$, $w<0$
      (c) $q<0$, $w>0$
      (d) $q>0$, $w<0$
    In case (b), $q<0$ and $w<0$ so ΔE is the most negative.

P18.2 $\Delta E = 0$ $q = p\Delta V = (14.0 \text{ atm})(12.0 \text{ L} - 1.0 \text{ L}) = 154 \text{ L atm}$

P18.3 $\Delta E = q - p\Delta V$ since $q = 0$
      $\Delta E = -p\Delta V$
    but ΔV is negative for a compression so ΔE increases and T increases.

P18.4 $\Delta E° = \Delta H° - \Delta nRT = -217.1 \text{ kJ} - (-1 \text{ mol})(8.314 \text{ J mol}^{-1} \text{ K}^{-1})(298 \text{ K})$
          $= -217.1 \text{ kJ} + 2.48 \text{ kJ}$
          $= -214.6 \text{ kJ}$

    % Difference $= (2.48/217) \times 100 = 1.14$ %

P18.5   (a) ΔS is negative since the products have a lower entropy, i.e. a lower freedom of movement.
      (b) ΔS is positive since the products have a higher entropy, i.e. a higher freedom of movement.

P18.6   (a)   There are more reactant molecules of gas than there are product molecules of gas, and we conclude that the entropy decreases, i.e. that ΔS is negative.
      (b)   Entropy decreases as this reaction proceeds because three moles of gaseous material disappear, and are replaced by only one mole of gaseous product. ΔS is negative.
      (c)   The sign of ΔS is negative, because three moles of gaseous reactants condense to a liquid product.
      (d)   The sign of ΔS is negative, because Δn is -2.
      (e)   The sign of ΔS is positive (entropy increases) because one mole of solid produces three moles of ions, and because a solid has dissolved and, therefore, been dispersed in a solvent.

P18.7 $\Delta S° = (\text{sum } S°[\text{products}]) - (\text{sum } S°[\text{reactants}])$

    (a) $\Delta S° = \{S°[H_2O(\ell)] + S°[CaCl_2(s)]\} - \{S°[CaO(s)] + 2S°[HCl(g)]\}$
          $\Delta S° = \{1 \text{ mol} \times (69.96 \text{ J mol}^{-1} \text{ K}^{-1}) + 1 \text{ mol} \times (114 \text{ J mol}^{-1} \text{ K}^{-1})\}$
              $- \{1 \text{ mol} \times (40 \text{ J mol}^{-1} \text{ K}^{-1}) + 2 \text{ mol} \times (186.7 \text{ J mol}^{-1} \text{ K}^{-1})\}$
          $\Delta S° = -229 \text{ J/K}$

    (b) $\Delta S° = \{S°[C_2H_6(g)]\} - \{S°[H_2(g)] + S°[C_2H_4(g)]\}$
          $\Delta S° = \{1 \text{ mol} \times (229.5 \text{ J mol}^{-1} \text{ K}^{-1})\}$
              $- \{1 \text{ mol} \times (130.6 \text{ J mol}^{-1} \text{ K}^{-1}) + 1 \text{ mol} \times (219.8 \text{ J mol}^{-1} \text{ K}^{-1})\}$
          $\Delta S° = -120.9 \text{ J/K}$

P18.8   First, we calculate $\Delta S°$, using the data:

    $\Delta S° = \{2S°[Fe_2O_3(s)]\} - \{3S°[O_2(g)] + 4S°[Fe(s)]\}$
    $\Delta S° = \{2 \text{ mol} \times (90.0 \text{ J mol}^{-1} \text{ K}^{-1})\}$
              $- \{3 \text{ mol} \times (205.0 \text{ J mol}^{-1} \text{ K}^{-1}) + 4 \text{ mol} \times (27 \text{ J mol}^{-1} \text{ K}^{-1})\}$
    $\Delta S° = -543 \text{ J/K} = -0.543 \text{ kJ/mol}$

Next, we calculate $\Delta H°$ using the data :

$\Delta H° = (\text{sum } \Delta H_f°[\text{products}]) - (\text{sum } \Delta H_f°[\text{reactants}])$
$\Delta H° = \{2\Delta H_f°[Fe_2O_3(s)]\} - \{3\Delta H_f°[O_2(g)] + 4\Delta H_f°[Fe(s)]\}$
$\Delta H° = \{2 \text{ mol} \times (-822.2 \text{ kJ/mol})\} - \{3 \text{ mol} \times (0.0 \text{ kJ/mol}) + 4 \times (0.0 \text{ kJ/mol})\}$
$\Delta H° = -1644 \text{ kJ}$

Now the temperature is 25.0 + 273.15 = 298.15 K, and the calculation of $\Delta G°$ is as follows: $\Delta G° = \Delta H° -$
$T\Delta S° = -1644 \text{ kJ} - (298.15 \text{ K})(-0.543 \text{ kJ/K}) = -1482 \text{ kJ}$

P18.9 $\Delta G° = (\text{sum } \Delta G_f°[\text{products}]) - (\text{sum } \Delta G_f°[\text{reactants}])$

(a) $\Delta G° = \{2\Delta G_f°[NO_2(g)]\} - \{\Delta G_f°[O_2(g)] + 2\Delta G_f°[NO(g)]\}$
           $\Delta G° = \{2 \text{ mol} \times (51.84 \text{ kJ/mol})\}$
                  $- \{1 \text{ mol} \times (0.0 \text{ kJ/mol}) + 2 \text{ mol} \times (86.69 \text{ kJ/mol})\}$
           $\Delta G° = -69.7 \text{ kJ}$

(b) $\Delta G° = \{2\Delta G_f°[H_2O(g)] + \Delta G_f°[CaCl_2(s)]\}$
                  $- \{2\Delta G_f°[HCl(g)] + \Delta G_f°[Ca(OH)_2(s)]\}$
           $\Delta G° = \{2 \text{ mol} \times (-228.6 \text{ kJ/mol}) + 1 \text{ mol} \times (-750.2 \text{ kJ/mol})\}$
                  $- \{2 \text{ mol} \times (-95.27 \text{ kJ/mol}) + 1 \text{ mol} \times (-896.76 \text{ kJ/mol})\}$
           $\Delta G° = -120.1 \text{ kJ}$

P18.10   The maximum amount of work that is available is the free energy change for the process, in this case, the standard free energy change, $\Delta G°$, since the process occurs at 25 °C.

$4Al(s) + 3O_2(g) \rightarrow 2Al_2O_3(s)$

$\Delta G° = (\text{sum } \Delta G_f°[\text{products}]) - (\text{sum } \Delta G_f°[\text{reactants}])$
$\Delta G° = 2\Delta G_f°[Al_2O_3(s)] - \{3\Delta G_f°[O_2(g)] + 4\Delta G_f°[Al(s)]\}$
$\Delta G° = 2 \text{ mol} \times (-1576.4 \text{ kJ/mol}) - \{3 \text{ mol} \times (0.0 \text{ kJ/mol}) + 4 \text{ mol} \times (0.0 \text{ kJ/mol})\}$
$\Delta G° = -3152.8 \text{ kJ}$, for the reaction as written.

This calculation conforms to the reaction as written. This means that the above value of $\Delta G°$ applies to the equation involving 4 mol of Al. The conversion to give energy per mole of aluminum is then: $-3152.8 \text{ kJ}/4$ mol Al = -788 kJ/mol

The maximum amount of energy that may be obtained is thus 788 kJ.

P18.11   For the vaporization process in particular, and for any process in general, we have:
$$\Delta G = \Delta H - T\Delta S$$
If the temperature is taken to be that at which equilibrium is obtained, that is the temperature of the boiling point (where liquid and vapor are in equilibrium with one another), then we also have the result that ΔG is equal to zero:
$$\Delta G = 0 = \Delta H - T\Delta S, \text{ or } T_{eq} = \Delta H/\Delta S$$
We know ΔH to be 60.7 kJ/mol, and we need the value for ΔS in units kJ mol^{-1} K^{-1}:

$\Delta S° = (\text{sum } S°[\text{products}]) - (\text{sum } S°[\text{reactants}])$
$\Delta S° = S°[Hg(g)] - S°[Hg(\ell)]$
$\Delta S° = (175 \times 10^{-3} \text{ kJ mol}^{-1} \text{ K}^{-1}) - (76.1 \times 10^{-3} \text{ kJ mol}^{-1} \text{ K}^{-1})$
$\Delta S° = 98.9 \times 10^{-3} \text{ kJ mol}^{-1} \text{ K}^{-1}$

$T_{eq} = 60.7 \text{ kJ/mol} \div 98.9 \times 10^{-3} \text{ kJ/mol K} = 614 \text{ K } (341 °C)$

P18.12 ΔG° = (sum ΔG_f°[products]) – (sum ΔG_f°[reactants])

$\Delta G^\circ = 2\Delta G_f^\circ[SO_3(g)] - \{2\Delta G_f^\circ[SO_2(g)] + \Delta G_f^\circ[O_2(g)]\}$

ΔG° = 2 mol × (–370.4 kJ/mol) – {2 mol × (–300.4 kJ/mol) + (0.0 kJ/mol)}

ΔG° = –140.0 kJ/mol

Since the sign of ΔG° is negative, the reaction should be spontaneous.

P18.13 $\Delta G^\circ = \Delta H^\circ - T\Delta S^\circ$

$\Delta G^\circ = \{2\Delta G_f^\circ[HCl(g)] + \Delta G_f^\circ[CaCO_3(s)]\}$
$\qquad\qquad - \{\Delta G_f^\circ[CaCl_2(s)] + \Delta G_f^\circ[H_2O(g)] + \Delta G_f^\circ[CO_2(g)]\}$

ΔG° = {2 mol × (–95.27 kJ/mol) + 1 mol × (–1128.8 kJ/mol)}
$\qquad$ – {1 mol × (–750.2 kJ/mol) + 1 mol × (–228.6 kJ/mol) + 1 mol × (–394.4 kJ/mol)}

ΔG° = +53.9 kJ

Since ΔG° is positive, the reaction is not spontaneous, and we do not expect to see products formed from reactants.

P18.14 First, we compute the standard free energy change for the reaction, based on the:

ΔG° = (sum ΔG_f°[products]) – (sum ΔG_f°[reactants])

$\Delta G^\circ = \{\Delta G_f^\circ[H_2O(g)] + \Delta G_f^\circ[CO_2(g)] + \Delta G_f^\circ[Na_2CO_3(s)]\} - \{2\Delta G_f^\circ[NaHCO_3(s)]\}$

ΔG° = {1 mol × (–228.6 kJ/mol) + 1 mol × (–394.4 kJ/mol)
$\qquad\qquad$ + 1 mol × (–1048 kJ/mol)} – {2 mol × (–851.9 kJ/mol)}

ΔG° = +33 kJ

Next, we determine values for ΔH° and ΔS°:

ΔH° = (sum ΔH_f°[products]) – (sum ΔH_f°[reactants])

$\Delta H^\circ = \{\Delta H_f^\circ[H_2O(g)] + \Delta H_f^\circ[CO_2(g)] + \Delta H_f^\circ[Na_2CO_3(s)]\} - \{2\Delta H_f^\circ[NaHCO_3(s)]\}$

ΔH° = {1 mol × (–241.8 kJ/mol) + 1 mol × (–393.5 kJ/mol)
$\qquad\qquad$ + 1 mol × (–1131 kJ/mol)} – {2 mol × (–947.7 kJ/mol)}

ΔH° = +129 kJ

ΔS° = (sum S°[products]) – (sum S°[reactants])

$\Delta S^\circ = \{S^\circ[H_2O(g)] + S^\circ[CO_2(g)] + S^\circ[Na_2CO_3(s)]\} - \{2S^\circ[NaHCO_3(s)]\}$

ΔS° = {1 mol × (188.7 J mol^{-1} K^{-1}) + 1 mol × (213.6 J mol^{-1} K^{-1})
$\qquad\qquad$ + 1 mol × (136 J mol^{-1} K^{-1})} – {2 mol × (102 J mol^{-1} K^{-1})}

ΔS° = 334 J/K = 0.334 kJ/K

Next, we assume that both ΔH° and ΔS° are independent of temperature, and use these values to determine ΔG at a temperature of 200 + 273 = 473 K:

$\Delta G^\circ_{473} = \Delta H^\circ - T\Delta S^\circ$ = 129 kJ – (473 K)(0.334 kJ/K) = –29 kJ

At the lower of these two temperatures (25 °C), the reaction has a positive value of ΔG. At the higher of these two temperatures (200 °C), the reaction has a negative value of ΔG. Thus ΔG becomes more negative as the temperature is raised, so the reaction becomes increasingly more favorable as the temperature is increased. In other words, the position of the equilibrium will be shifted more towards products at the higher temperature.

P18.15   Using the data provided we may write:

$$\Delta G = \Delta G° + RT \ln\left(\frac{P_{N_2O_4}}{P_{NO_2}^2}\right)$$

$$= -5.40 \times 10^3 \text{ J mol}^{-1} + \left(8.314 \text{ J mol}^{-1} \text{ K}^{-1}\right)\left(298 \text{ K}\right)\ln\left(\frac{0.25 \text{ atm}}{(0.60 \text{ atm})^2}\right)$$

$$= -5.40 \times 10^3 \text{ J mol}^{-1} + \left(-9.03 \times 10^2 \text{ J mol}^{-1}\right)$$

$$= -6.30 \times 10^3 \text{ J mol}^{-1}$$

Since ΔG is negative, the forward reaction is spontaneous and the reaction will proceed to the right.

P18.16 $\Delta G° = -RT \ln K_p$
$\Delta G° = -(8.314 \text{ J K}^{-1} \text{ mol}^{-1})(25 + 273 \text{ K}) \times \ln(6.9 \times 10^5) = -33.3 \times 10^3 \text{ J}$
$\Delta G° = -33.3 \text{ kJ}$

P18.17 $\Delta G° = -RT \ln K_p$
$3.3 \times 10^3 \text{ J} = -(8.314 \text{ J K}^{-1} \text{ mol}^{-1})(298 \text{ K}) \times \ln(K_p)$
$\ln(K_p) = -3.3 \times 10^3 \text{ J}/[(8.314 \text{ J K}^{-1} \text{ mol}^{-1})(298 \text{ K})] = -1.3$
Taking the antilog of both sides of the above equation gives: $K_p = 0.26$

P18.18 $\Delta G° = \Delta H° - T\Delta S° = -92.4 \times 10^3 \text{ J} - (323 \text{ K})(-198.3 \text{ J/K}) = -2.83 \times 10^4 \text{ J}$
$\Delta G = -RT \ln K$
Thus, $-2.83 \times 10^4 \text{ J} = -(8.314 \text{ J K}^{-1} \text{ mol}^{-1})(323 \text{ K}) \times \ln(K_p)$
$\ln(K_p) = -2.38 \times 10^4 \text{ J}/[(-8.314 \text{ J K}^{-1} \text{ mol}^{-1})(323 \text{ K})] = 10.5$
Taking the antilog of both sides of the above equation gives: $K_p = 3.84 \times 10^4$

Review Problems

18.50 $\Delta E = q + w = 300 \text{ J} + 700 \text{ J} = +1000 \text{ J}$

The overall process is endothermic, meaning that the internal energy of the system increases. Notice that both terms, q and w, contribute to the increase in internal energy of the system; the system gains heat (+q) and has work done on it (+w).

18.52   work = $P \times \Delta V$
The total pressure is atmospheric pressure plus that caused by the hand pump:
$P = (30.0 + 14.7) \text{ lb/in}^2 = 44.7 \text{ lb/in}^2$

Converting to atmospheres we get:
$P = 44.7 \text{ lb/in}^2 \times 1 \text{ atm}/14.7 \text{ lb/in}^2 = 3.04 \text{ atm}$

Next we convert the volume change in units in^3 to units L:
$24.0 \text{ in}^3 \times (2.54 \text{ cm/in})^3 \times 1 \text{ L}/1000 \text{ cm}^3 = 0.393 \text{ L}$

Hence $P \times \Delta V = (3.04 \text{ atm})(0.393 \text{ L}) = 1.19 \text{ L•atm}$
$\qquad\qquad 1.19 \text{ L•atm} \times 101.3 \text{ J/L•atm} = 121 \text{ J}$

18.54 We use the data supplied in Appendix.

(a) $3PbO(s) + 2NH_3(g) \rightarrow 3Pb(s) + N_2(g) + 3H_2O(g)$

$\Delta H° = \{3\Delta H_f°[Pb(s)] + \Delta H_f°[N_2(g)] + 3\Delta H_f°[H_2O(g)]\}$
$\quad\quad\quad\quad - \{3\Delta H_f°[PbO(s)] + 2\Delta H_f°NH_3(g)]\}$

$\Delta H° = \{3 \text{ mol} \times (0 \text{ kJ/mol}) + 1 \text{ mol} \times (0 \text{ kJ/mol}) + 3 \text{ mol} \times (-241.8 \text{ kJ/mol})\}$
$\quad\quad\quad\quad - \{3 \text{ mol} \times (-217.3 \text{ kJ/mol}) + 2 \text{ mol} \times (-46.19 \text{ kJ/mol})\}$

$\Delta H° = + 24.58 \text{ kJ}$

$\Delta E = \Delta H° - \Delta nRT$
$\Delta E = 24.58 \text{ kJ} - (+2 \text{ mol})(8.314 \text{ J/mol K})(10^{-3} \text{ kJ/J})(298 \text{ K}) = 19.6 \text{ kJ}$

(b) $NaOH(s) + HCl(g) \rightarrow NaCl(s) + H_2O(\ell)$

$\Delta H° = \{\Delta H_f°[NaCl(s)] + \Delta H_f°[H_2O(\ell)]\} - \{\Delta H_f°[NaOH(s)] + \Delta H_f°[HCl(g)]\}$
$\Delta H° = \{1 \text{ mol} \times (-411.0 \text{ kJ/mol}) + 1 \text{ mol} \times (-285.9 \text{ kJ/mol})\}$
$\quad\quad\quad\quad - \{1 \text{ mol} \times (-426.8 \text{ kJ/mol}) + 1 \text{ mol} \times (-92.30)\}$

$\Delta H° = -178 \text{ kJ}$

$\Delta E = \Delta H° - \Delta nRT$
$\Delta E = -178 \text{ kJ} - (-1)(8.314 \text{ J/mol K})(10^{-3} \text{ kJ/J})(298 \text{ K}) = -175 \text{ kJ}$

(c) $Al_2O_3(s) + 2Fe(s) \rightarrow Fe_2O_3(s) + 2Al(s)$

$\Delta H° = \{\Delta H_f°[Fe_2O_3(s)] + 2\Delta H_f°[Al(s)]\} - \{\Delta H_f°[Al_2O_3(s)] + 2\Delta H_f°[Fe(s)]\}$
$\Delta H° = \{1 \text{ mol} \times (-822.3 \text{ kJ/mol}) + 2 \text{ mol} \times (\text{o kJ/mol})\}$
$\quad\quad\quad\quad - \{1 \text{ mol} \times (-1669.8 \text{ kJ/mol}) + 2 \text{ mol} \times (0 \text{ kJ/mol})\}$

$\Delta H° = 847.6 \text{ kJ}$

$\Delta E = \Delta H°$, since the value of Δn for this reaction is zero.

(d) $2CH_4(g) \rightarrow C_2H_6(g) + H_2(g)$
$\Delta H° = \{\Delta H_f°[C_2H_6(g)] + \Delta H_f°[H_2(g)]\} - \{2\Delta H_f°[CH_4(g)]\}$
$\Delta H° = \{1 \text{ mol} \times (-84.667 \text{ kJ/mol}) + 1 \text{ mol} \times (0.0 \text{ kJ/mol})\}$
$\quad\quad\quad\quad - \{2 \text{ mol} \times (-74.848 \text{ kJ/mol})\}$

$H° = 65.029 \text{ kJ}$

$\Delta E = \Delta H°$, since the value of Δn for this reaction is zero.

18.56 In general, we have the equation: $\Delta H° = (\text{sum } \Delta H_f°[\text{products}]) - (\text{sum } \Delta H_f°[\text{reactants}])$

(a) $\Delta H° = \{\Delta H_f°[CaCO_3(s)]\} - \{\Delta H_f°[CO_2(g)] + \Delta H_f°[CaO(s)]\}$
$\Delta H° = \{1 \text{ mol} \times (-1207 \text{ kJ/mol})\}$
$\quad\quad\quad - \{1 \text{ mol} \times (-394 \text{ kJ/mol}) + 1 \text{ mol} \times (-635.5 \text{ kJ/mol})\}$
$\Delta H° = -178 \text{ kJ} \quad \therefore \text{ favored.}$

(b) $\Delta H° = \{\Delta H_f°[C_2H_6(g)]\} - \{\Delta H_f°[C_2H_2(g)] + 2\Delta H_f°[H_2(g)]\}$
$\Delta H° = \{1 \text{ mol} \times (-84.5 \text{ kJ/mol})\}$
$\quad\quad\quad\quad - \{1 \text{ mol} \times (227 \text{ kJ/mol}) + 2 \text{ mol} \times (0.0 \text{ kJ/mol})\}$
$\Delta H° = -311 \text{ kJ} \quad \therefore \text{ favored.}$

(c) $\Delta H° = \{\Delta H_f°[Fe_2O_3(s)] + 3\Delta H_f°[Ca(s)]\}$
$\qquad - \{2\Delta H_f°[Fe(s)] + 3\Delta H_f°[CaO(s)]\}$

$\Delta H° = \{1 \text{ mol} \times (-822.2 \text{ kJ/mol}) + 3 \text{ mol} \times (0.0 \text{ kJ/mol})\}$
$\qquad - \{2 \text{ mol} \times (0.0 \text{ kJ/mol}) + 3 \text{ mol} \times (-635.5 \text{ kJ/mol})\}$

$\Delta H° = +1084.3 \text{ kJ} \;\therefore\; \text{not favorable from the standpoint of enthalpy alone.}$

(d) $\Delta H° = \{\Delta H_f°[H_2O(\ell)] + \Delta H_f°[CaO(s)]\} - \{\Delta H_f°[Ca(OH)_2(s)]\}$

$\Delta H° = \{1 \text{ mol} \times (-285.9 \text{ kJ/mol}) + 1 \text{ mol} \times (-635.5 \text{ kJ/mol})\}$
$\qquad - \{1 \text{ mol} \times (-986.59 \text{ kJ/mol})\}$

$\Delta H° = +65.2 \text{ kJ} \;\therefore\; \text{not favored from the standpoint of enthalpy alone.}$

(e) $\Delta H° = \{2\Delta H_f°[HCl(g)] + \Delta H_f°[Na_2SO_4(s)]\}$
$\qquad - \{2\Delta H_f°[NaCl(s)] + \Delta H_f°[H_2SO_4(\ell)]\}$

$\Delta H° = \{2 \text{ mol} \times (-92.30 \text{ kJ/mol}) + 1 \text{ mol} \times (-1384.5 \text{ kJ/mol})\}$
$\qquad - \{2 \text{ mol} \times (-411.0 \text{ kJ/mol}) + 1 \text{ mol} \times (-811.32 \text{ kJ/mol})\}$

$\Delta H° = +64.2 \text{ kJ} \;\therefore\; \text{not favored from the standpoint of enthalpy alone.}$

18.58 The probability is given by the number of possibilities that lead to the desired arrangement, divided by the total number of possible arrangements.

We list each of the possible results, heads (H) or tails (T) for each of the coins, and systematically write down all of the distinct arrangements:

There is only one arrangement that gives four heads: HHHH.
There is only one arrangement that gives four tails: TTTT
Four distinct arrangements can lead to three heads and one tail:
    HHHT, THHH, HTHH, HHTH
There are similarly four distinct arrangements that lead to three tails and one head:
    TTTH, HTTT, THTT, TTHT
There are six distinct arrangements that can lead to two heads and two tails:
    HHTT, TTHH, THHT, THTH, HTTH, HTHT

Hence, the probability of all heads (HHHH) is 1 in 16 or 1/16 = 0.0625.
The probability of two heads and two tails is 6 in 16 or 6/16 = 0.375.

18.60 (a) negative – since the number of moles of gaseous material decreases.
(b) negative – since the number of moles of gaseous material decreases.
(c) negative – since the number of moles of gas decreases.
(d) positive – since a gas appears where there formerly was none.

18.62 $\Delta S° = (\text{sum } S°[\text{products}]) - (\text{sum } S°[\text{reactants}])$

(a) $\Delta S° = \{2S°[NH_3(g)]\} - \{3S°[H_2(g)] + S°[N_2(g)]\}$
$\Delta S° = \{2 \text{ mol} \times (192.5 \text{ J mol}^{-1} \text{ K}^{-1})\} - \{3 \text{ mol} \times (130.6 \text{ J mol}^{-1} \text{ K}^{-1})$
$\qquad + 1 \text{ mol} \times (191.5 \text{ J mol}^{-1} \text{ K}^{-1})\}$

$\Delta S° = -198.3 \text{ J/K} \;\therefore\; \text{not spontaneous from the standpoint of entropy.}$

(b) $\Delta S° = \{S°[CH_3OH(\ell)]\} - \{2S°[H_2(g)] + S°[CO(g)]\}$
$\Delta S° = \{1 \text{ mol} \times (126.8 \text{ J mol}^{-1} \text{ K}^{-1})\}$
$\qquad - \{2 \text{ mol} \times (130.6 \text{ J mol}^{-1} \text{ K}^{-1}) + 1 \text{ mol} \times (197.9 \text{ J mol}^{-1} \text{ K}^{-1})\}$

$\Delta S° = -332.3 \text{ J/K} \;\therefore\; \text{not favored from the standpoint of entropy alone.}$

(c) $\Delta S° = \{6S°[H_2O(g)] + 4S°[CO_2(g)]\} - \{7S°[O_2(g)] + 2S°[C_2H_6(g)]\}$
$\Delta S° = \{6 \text{ mol} \times (188.7 \text{ J mol}^{-1} \text{ K}^{-1}) + 4 \text{ mol} \times (213.6 \text{ J mol}^{-1} \text{ K}^{-1})\}$
$\qquad - \{7 \text{ mol} \times (205.0 \text{ J mol}^{-1} \text{ K}^{-1}) + 2 \text{ mol} \times (229.5 \text{ J mol}^{-1} \text{ K}^{-1})\}$
$\Delta S° = +92.6 \text{ J/K} \therefore$ favorable from the standpoint of entropy alone.

(d) $\Delta S° = \{2S°[H_2O(\ell)] + S°[CaSO_4(s)]\}$
$\qquad\qquad - \{S°[H_2SO_4(\ell)] + S°[Ca(OH)_2(s)]\}$
$\Delta S° = \{2 \text{ mol} \times (69.96 \text{ J mol}^{-1} \text{ K}^{-1}) + 1 \text{ mol} \times (107 \text{ J mol}^{-1} \text{ K}^{-1})\}$
$\qquad\qquad - \{1 \text{ mol} \times (157 \text{ J mol}^{-1} \text{ K}^{-1}) + 1 \text{ mol} \times (76.1 \text{ J mol}^{-1} \text{ K}^{-1})\}$
$\Delta S° = +14 \text{ J/K} \therefore$ favorable from the standpoint of entropy alone.

(e) $\Delta S° = \{2S°[N_2(g)] + S°[SO_2(g)]\} - \{2S°[N_2O(g)] + S°[S(s)]\}$
$\Delta S° = \{2 \text{ mol} \times (191.5 \text{ J mol}^{-1} \text{ K}^{-1}) + 1 \text{ mol} \times (248 \text{ J mol}^{-1} \text{ K}^{-1})\}$
$\qquad\qquad - \{2 \text{ mol} \times (220.0 \text{ J mol}^{-1} \text{ K}^{-1}) + 1 \text{ mol} \times (31.8 \text{ J mol}^{-1} \text{ K}^{-1})\}$
$\Delta S° = +159 \text{ J/K} \therefore$ favorable from the standpoint of entropy alone.

18.64 The entropy change that is designated $\Delta S_f°$ is that which corresponds to the reaction in which one mole of a substance is formed from elements in their standard states. Since the value is understood to correspond to the reaction forming one mole of a single pure substance, the units may be written either J K^{-1} or $\text{J mol}^{-1} \text{ K}^{-1}$.

(a) $2C(s) + 2H_2(g) \rightarrow C_2H_4(g)$
$\Delta S° = \{S°[C_2H_4(g)]\} - \{2S°[C(s)] + 2S°[H_2(g)]\}$
$\Delta S° = \{1 \text{ mol} \times (219.8 \text{ J mol}^{-1} \text{ K}^{-1})\}$
$\qquad\qquad - \{2 \text{ mol} \times (5.69 \text{ J mol}^{-1} \text{ K}^{-1}) + 2 \text{ mol} \times (130.6 \text{ J mol}^{-1} \text{ K}^{-1})\}$
$\Delta S° = -52.8 \text{ J/K or } -52.8 \text{ J mol}^{-1} \text{ K}^{-1}$

(b) $N_2(g) + 1/2O_2(g) \rightarrow N_2O(g)$
$\Delta S° = \{S°[N_2O(g)]\} - \{S°[N_2(g)] + 1/2S°[O_2(g)]\}$
$\Delta S° = \{1 \text{ mol} \times (220.0 \text{ J mol}^{-1} \text{ K}^{-1})\} - \{1 \text{ mol} \times (191.5 \text{ J mol}^{-1} \text{ K}^{-1})$
$\qquad\qquad + 1/2 \text{ mol} \times (205.0 \text{ J mol}^{-1} \text{ K}^{-1})\}$
$\Delta S° = -74.0 \text{ J/K or } -74.0 \text{ J mol}^{-1} \text{ K}^{-1}$

(c) $Na(s) + 1/2Cl_2(g) \rightarrow NaCl(s)$
$\Delta S° = \{S°[NaCl(s)]\} - \{1/2S°[Cl_2(g)] + S°[Na(s)]\}$
$\Delta S° = \{1 \text{ mol} \times (72.38 \text{ J mol}^{-1} \text{ K}^{-1})\} - \{1/2 \text{ mol} \times 223.0 \text{ J mol}^{-1} \text{ K}^{-1})$
$\qquad\qquad + 1 \text{ mol} \times (51.0 \text{ J mol}^{-1} \text{ K}^{-1})\}$
$\Delta S° = -90.1 \text{ J/K or } -90.1 \text{ J mol}^{-1} \text{ K}^{-1}$

(d) $Ca(s) + S(s) + 3O_2(g) + 2H_2(g) \rightarrow CaSO_4 \bullet 2H_2O(s)$
$\Delta S° = \{S°[CaSO_4 \bullet 2H_2O(s)]\} - \{2S°[H_2(g)] + 3S°[O_2(g)] + S°[S(s)]$
$\qquad\qquad\qquad + S°[Ca(s)]\}$
$\Delta S° = \{1 \text{ mol} \times (194.0 \text{ J mol}^{-1} \text{ K}^{-1})\} - \{2 \text{ mol} \times (130.6 \text{ J mol}^{-1} \text{ K}^{-1})$
$\qquad\qquad + 3 \text{ mol} \times (205.0 \text{ J mol}^{-1} \text{ K}^{-1}) + 1 \text{ mol} \times (31.8 \text{ J mol}^{-1} \text{ K}^{-1})$
$\qquad\qquad + 1 \text{ mol} \times (154.8 \text{ J mol}^{-1} \text{ K}^{-1})\}$
$\Delta S° = -868.9 \text{ J/K or } -868.9 \text{ J mol}^{-1} \text{ K}^{-1}$

(e) $2H_2(g) + 2C(s) + O_2(g) \rightarrow HC_2H_3O_2(\ell)$
$\Delta S° = \{S°[HC_2H_3O_2(\ell)]\} - \{2S°[H_2(g)] + 2S°[C(s)] + S°[O_2(g)]\}$
$\Delta S° = \{1 \text{ mol} \times (160 \text{ J mol}^{-1} \text{ K}^{-1})\} - \{2 \text{ mol} \times (130.6 \text{ J mol}^{-1} \text{ K}^{-1})$
$\qquad\qquad + 2 \text{ mol} \times (5.69 \text{ J mol}^{-1} \text{ K}^{-1}) + 1 \text{ mol} \times (205.0 \text{ J mol}^{-1} \text{ K}^{-1})\}$
$\Delta S° = -318 \text{ J/K or } -318 \text{ J mol}^{-1} \text{ K}^{-1}$

18.66 $\Delta S° = (sum \ S°[products]) - (sum \ S°[reactants])$

$\Delta S° = \{2S°[HNO_3(\ell)] + S°[NO(g)]\} - \{3S°[NO_2(g)] + S°[H_2O(\ell)]\}$

$\Delta S° = \{2 \ mol \times (155.6 \ J \ mol^{-1} \ K^{-1}) + 1 \ mol \times (210.6 \ J \ mol^{-1} \ K^{-1})\}$
$- \{3 \ mol \times (240.5 \ J \ mol^{-1} \ K^{-1}) + 1 \ mol \times (69.96 \ J \ mol^{-1} \ K^{-1})\}$

$\Delta S° = -269.7 \ J/K$

18.68   The quantity $\Delta G_f°$ applies to the equation in which one mole of pure phosgene is produced from the naturally occurring forms of the elements:

$$C(s) + 1/2O_2(g) + Cl_2(g) \rightarrow COCl_2(g), \ \Delta G_f° = ?$$

We can determine $\Delta G_f°$ if we can find values for $\Delta H_f°$ and $\Delta S_f°$, because:

$$\Delta G° = \Delta H° - T\Delta S°$$

The value of $\Delta S_f°$ is determined using S° for phosgene in the following way:

$\Delta S_f° = \{S°[COCl_2(g)]\} - \{S°[C(s)] + 1/2S°[O_2(g)] + S°[Cl_2(g)]\}$

$\Delta S_f° = \{1 \ mol \times (284 \ J \ mol^{-1} \ K^{-1})\} - \{1 \ mol \times (5.69 \ J \ mol^{-1} \ K^{-1})$
$+ 1/2 \ mol \times (205.0 \ J \ mol^{-1} \ K^{-1}) + 1 \ mol \times (223.0 \ J \ mol^{-1} \ K^{-1})\}$

$\Delta S_f° = -47 \ J \ mol^{-1} \ K^{-1} \ or \ -47 \ J/K$

$\Delta G_f° = \Delta H_f° - T\Delta S_f° = -223 \ kJ/mol - (298 \ K)(-0.047 \ kJ/mol \ K)$
$= -209 \ kJ/mol$

18.70 $\Delta G° = (sum \ \Delta G_f°[products]) - (sum \ \Delta G_f°[reactants])$

(a) $\Delta G° = \{\Delta G_f°[H_2SO_4(\ell)]\} - \{\Delta G_f°[H_2O(\ell)] + \Delta G_f°[SO_3(g)]\}$

$\Delta G° = \{1 \ mol \times (-689.9 \ kJ/mol)\} - \{1 \ mol \times (-237.2 \ kJ/mol)$
$+ 1 \ mol \times (-370 \ kJ/mol)\}$

$\Delta G° = -83 \ kJ$

(b) $\Delta G° = \{2\Delta G_f°[NH_3(g)] + \Delta G_f°[H_2O(\ell)] + \Delta G_f°[CaCl_2(s)]\}$
$- \{\Delta G_f°[CaO(s)] + 2\Delta G_f°[NH_4Cl(s)]\}$

$\Delta G° = \{2 \ mol \times (-16.7 \ kJ/mol) + 1 \ mol \times (-237.2 \ kJ/mol)$
$+ 1 \ mol \times (-750.2 \ kJ/mol)\} - \{1 \ mol \times (-604.2 \ kJ/mol)$
$+ 2 \ mol \times (-203.9 \ kJ/mol)\}$

$\Delta G° = -8.8 \ kJ$

(c) $\Delta G° = \{\Delta G_f°[H_2SO_4(\ell)] + \Delta G_f°[CaCl_2(s)]\} - \{\Delta G_f°[CaSO_4(s)]$
$+ 2\Delta G_f°[HCl(g)]\}$

$\Delta G° = \{1 \ mol \times (-689.9 \ kJ/mol) + 1 \ mol \times (-750.2 \ kJ/mol)\}$
$- \{1 \ mol \times (-1320.3 \ kJ/mol) + 2 \ mol \times (-95.27 \ kJ/mol)\}$

$\Delta G° = +70.7 \ kJ$

(d) $\Delta G° = \{\Delta G_f°[C_2H_5OH(\ell)]\} - \{\Delta G_f°[H_2O(g)] + \Delta G_f°[C_2H_4(g)]\}$

$\Delta G° = \{1 \ mol \times (-174.8 \ kJ/mol)\} - \{1 \ mol \times (-228.6 \ kJ/mol)$
$+ 1 \ mol \times (68.12 \ kJ/mol)\}$

$\Delta G° = -14.3 \ kJ$

(e) $\Delta G° = \{2\Delta G_f°[H_2O(\ell)] + \Delta G_f°[SO_2(g)] + \Delta G_f°[CaSO_4(s)]\}$
$\quad\quad\quad - \{2\Delta G_f°[H_2SO_4(\ell)] + \Delta G_f°[Ca(s)]\}$

$\Delta G° = \{2\text{ mol} \times (-237.2 \text{ kJ/mol}) + 1 \text{ mol} \times (-300 \text{ kJ/mol})$
$\quad\quad\quad + 1 \text{ mol} \times (-1320.3 \text{ kJ/mol})\} - \{2 \text{ mol} \times (-689.9 \text{ kJ/mol})$
$\quad\quad\quad + 1 \text{ mol} \times (0.0 \text{ kJ/mol})\}$

$\Delta G° = -715 \text{ kJ}$

18.72 $CaSO_4 \bullet 1/2H_2O(s) + 3/2H_2O(\ell) \rightarrow CaSO_4 \bullet 2H_2O(s)$
$\Delta G° = (\text{sum } \Delta G_f°[\text{products}]) - (\text{sum } \Delta G_f°[\text{reactants}])$
$\Delta G° = \{\Delta G_f°[CaSO_4 \bullet 2H_2O(s)]\} -$
$\quad\quad\quad\quad \{\Delta G_f°[CaSO_4 \bullet 1/2H_2O(s)] + 3/2\Delta G_f°[H_2O(\ell)]\}$
$\Delta G° = \{1 \text{ mol} \times (-1795.7 \text{ kJ/mol})\}$
$\quad\quad\quad - \{1 \text{ mol} \times (-1435.2 \text{ kJ/mol}) + 1.5 \text{ mol} \times (-237.2 \text{ kJ/mol})\}$
$\Delta G° = -4.7 \text{ kJ}$

18.74   Multiply the reverse of the second equation by 2 (remembering to multiply the associated free energy change by −2), and add the result to the first equation:

$4NO(g) \rightarrow 2N_2O(g) + O_2(g),$ $\quad\quad\quad\quad \Delta G° = -139.56 \text{ kJ}$
$4NO_2(g) \rightarrow 4NO(g) + 2O_2(g),$ $\quad\quad\quad\quad \Delta G° = +139.40 \text{ kJ}$

$4NO_2 \rightarrow 3O_2(g) + 2N_2O(g),$ $\quad\quad\quad\quad \Delta G° = -0.16 \text{ kJ}$

This result is the reverse of the desired reaction, which must then have $\Delta G° = +0.16$ kJ

18.76   The maximum work obtainable from a reaction is equal in magnitude to the value of ΔG for the reaction. Thus, we need only determine $\Delta G°$ for the process:

$\Delta G° = (\text{sum } \Delta G_f°[\text{products}]) - (\text{sum } \Delta G_f°[\text{reactants}])$
$\Delta G° = \{3\Delta G_f°[H_2O(g)] + 2\Delta G_f°[CO_2(g)]\} - \{3\Delta G_f°[O_2(g)] + \Delta G_f°[C_2H_5OH(\ell)]\}$
$\Delta G° = \{3 \text{ mol} \times (-228.6 \text{ kJ/mol}) + 2 \text{ mol} \times (-394.4 \text{ kJ/mol})\}$
$\quad\quad\quad\quad - \{3 \text{ mol} \times (0.0 \text{ kJ/mol}) \ 1 \text{ mol} \times (-174.8 \text{ kJ/mol})\}$
$\Delta G° = -1299.8 \text{ kJ}$

18.78   At equilibrium, $\Delta G = 0 = \Delta H - T\Delta S$
$T_{eq} = \Delta H/\Delta S$, and assuming that ΔS is independent of temperature, we have:
$T_{eq} = (31.4 \times 10^3 \text{ J mol}^{-1}) \div (94.2 \text{ J mol}^{-1} \text{ K}^{-1}) = 333 \text{ K}$

18.80   At equilibrium, $\Delta G = 0 = \Delta H - T\Delta S$
Thus $\Delta H = T\Delta S$, and if we assume that both ΔH and ΔS are independent of temperature, we have:

$\Delta S = \Delta H/T_{eq} = (37.7 \times 10^3 \text{ J/mol}) \div (99.3 + 273.15 \text{ K})$
$\Delta S = 101.2 \text{ J mol}^{-1} \text{ K}^{-1}$

18.82   The reaction is spontaneous if its associated value for $\Delta G°$ is negative.
$\Delta G° = (\text{sum } \Delta G_f°[\text{products}]) - (\text{sum } \Delta G_f°[\text{reactants}])$
$\Delta G° = \{\Delta G_f°[HC_2H_3O_2(\ell)] + \Delta G_f°[H_2O(\ell)] + \Delta G_f°[NO(g)] + \Delta G_f°[NO_2(g)]\}$
$\quad\quad\quad - \{\Delta G_f°[C_2H_4(g)] + 2\Delta G_f°[HNO_3(\ell)]\}$

$\Delta G° = \{1 \text{ mol} \times (-392.5 \text{ kJ/mol}) + 1 \text{ mol} \times (-237.2 \text{ kJ/mol})$
$\qquad\qquad + 1 \text{ mol} \times (86.69 \text{ kJ/mol}) + 1 \text{ mol} \times (51.84 \text{ kJ/mol})\}$
$\qquad\qquad - \{1 \text{ mol} \times (68.12 \text{ kJ/mol}) + 2 \text{ mol} \times (-79.9 \text{ kJ/mol})\}$

$\Delta G° = -399.5 \text{ kJ}$

Yes, the reaction is spontaneous.

18.84 $\Delta G°_T = \Delta H° - T\Delta S°$, where T = 373 K in all cases, and where the values of $\Delta H°$ and $\Delta S°$ are obtained in the usual manner, i.e.:

$\Delta H° = (\text{sum } \Delta H_f°[\text{products}]) - (\text{sum } \Delta H_f°[\text{reactants}])$
$\Delta S° = (\text{sum } S°[\text{products}]) - (\text{sum } S°[\text{reactants}])$

$\Delta H° = \{\Delta H_f°[C_2H_6(g)]\} - \{\Delta H_f°[H_2(g)] + \Delta H_f°[C_2H_4(g)]\}$
$\Delta H° = \{1 \text{ mol} \times (-84.5 \text{ kJ/mol})\}$
$\qquad\qquad - \{1 \text{ mol} \times (0.0 \text{ kJ/mol}) + 1 \text{ mol} \times (51.9 \text{ kJ/mol})\}$
$\Delta H° = -136.4 \text{ kJ}$

$\Delta S° = \{S°[C_2H_6(g)]\} - \{S°[H_2(g)] + S°[C_2H_4(g)]\}$
$\Delta S° = \{1 \text{ mol} \times (229.5 \text{ J mol}^{-1} \text{ K}^{-1})\} - \{1 \text{ mol} \times (130.6 \text{ J mol}^{-1} \text{ K}^{-1})$
$\qquad\qquad + 1 \text{ mol} \times (219.8 \text{ J mol}^{-1} \text{ K}^{-1})\}$
$\Delta S° = -120.9 \text{ J/K} = -0.1209 \text{ kJ/K}$

$\Delta G°_{373} = \Delta H° - T\Delta S° = -136.4 \text{ kJ} - (373 \text{ K})(-0.1209 \text{ kJ/K}) = -91.3 \text{ kJ}$

18.86 (a) $\Delta G° = \{2 \times \Delta G_f°[POCl_3(g)]\} - \{2 \times \Delta G_f°[PCl_3(g)] + 2 \times \Delta G_f°[O_2(g)]\}$
$\Delta G° = \{2 \text{ mol} \times (-1019 \text{ kJ/mol})\}$
$\qquad\qquad - \{2 \text{ mol} \times (-267.8 \text{ kJ/mol}) + 1 \text{ mol} \times (0 \text{ kJ/mol})\}$
$\Delta G° = -1502 \text{ kJ} = -1.502 \times 10^6 \text{ J}$
$-1.502 \times 10^6 \text{ J} = -RT\ln K_p = -(8.314 \text{ J/K mol})(298 \text{ K}) \times \ln K_p$
$\ln K_p = 606 \quad \therefore \log K_p = 263$, and $K_p = 10^{263}$.

(b) $\Delta G° = \{2 \times \Delta G_f°[SO_2(g)] + 1 \times \Delta G_f°[O_2(g)]\} - \{2 \times \Delta G_f°[SO_3(g)]\}$
$\Delta G° = \{2 \text{ mol} \times (-300 \text{ kJ/mol}) + 1 \text{ mol} \times (0 \text{ kJ/mol})\} - \{2 \text{ mol} \times (-370 \text{ kJ/mol})\}$
$\Delta G° = 140 \text{ kJ} = 1.40 \times 10^5 \text{ J}$
$1.40 \times 10^5 \text{ J} = -RT\ln K_p = -(8.314 \text{ J/K mol})(298 \text{ K}) \times \ln K_p$
$\ln K_p = -56.5$ and $K_p = 2.90 \times 10^{-25}$

18.88 $\Delta G° = -RT \ln K_p$
$-9.67 \times 10^3 \text{ J} = -(8.314 \text{ J/K mol})(1273 \text{ K}) \times \ln K_p$
$\ln K_p = 0.914 \quad \therefore K_p = 2.49$

$$Q = \frac{[N_2O][O_2]}{[NO_2][NO]} = \frac{(0.015)(0.0350)}{(0.0200)(0.040)} = 0.66$$

Since the value of Q is less than the value of K, the system is not at equilibrium and must shift to the right to reach equilibrium.

18.90 $\Delta G° = -RT \ln K_p$
$-50.79 \times 10^3 \text{ J} = -(8.314 \text{ J K}^{-1} \text{ mol}^{-1})(298 \text{ K}) \times \ln K_p$
$\ln K_p = 20.50$
Taking the antiln of both sides of this equation gives: $K_p = 8.000 \times 10^8$

This is a favorable reaction, since the equilibrium lies far to the side favoring products and is worth studying as a method for methane production.

18.92   If $\Delta G° = 0$, $K_c = 1$. If we start with pure products, the value of Q will be infinite (there are zero reactants) and, since $Q > K_c$, the equilibrium will shift towards the reactants, i.e., the pure products will decompose to their elements.

18.94 $\Delta G° = (\text{sum } \Delta G_f°[\text{products}]) - (\text{sum } \Delta G_f°[\text{reactants}])$
$\Delta G° = \{\Delta G_f°[N_2(g)] + 2\Delta G_f°[CO_2(g)]\} - \{2\Delta G_f°[NO(g)] + 2\Delta G_f°[CO(g)]\}$
$\Delta G° = \{1 \text{ mol} \times (0.0 \text{ kJ/mol}) + 2 \text{ mol} \times (-394.4 \text{ kJ/mol})\}$
$\qquad\qquad\qquad - \{2 \text{ mol} \times (86.69 \text{ kJ/mol}) + 2 \text{ mol} \times (-137.3 \text{ kJ/mol})\}$
$\Delta G° = -687.6 \text{ kJ}$

$\Delta G° = -RT \ln K_p$
$-687.6 \times 10^3 \text{ J} = -(8.314 \text{ J K}^{-1} \text{ mol}^{-1})(298 \text{ K}) \ln K_p$
$\ln K_p = 278$ and $\log K_p = 121$ $\therefore K_p = 10^{121}$

18.96   This requires the breaking of three N–H single bonds:

$$NH_3 \rightarrow N + 3H$$

The enthalpy of atomization of NH_3 is thus three times the average N–H single bond energy: 3×391 kJ/mol $= 1.17 \times 10^3$ kJ/mol

18.98   We proceed as in the answer to review exercise 18.94. The various bonds that compose the molecule are:
     5 C—H bonds
     1 C—C bond
     1 C—O bond
     1 O—H bond

$\Delta H_f° = \text{sum}(\Delta H_f°[\text{gaseous atoms}]) - \text{sum}(\text{average bond energies in the molecule})$
$\Delta H_f°[C_2H_5OH(g)] = -235.3$ kJ/mol

$\qquad = [2 \times 716.67 + 6 \times 217.89 + 249.17] - [5 \times 412 + 348 + 463 + \text{C—O}]$

from which we can calculate that the C—O bond energy is 354 kJ/mol.

18.100   There are two C=S double bonds to be considered:
$\Delta H_f° = \text{sum}(\Delta H_f°[\text{gaseous atoms}]) - \text{sum}(\text{average bond energies in the molecule})$
$\Delta H_f°[CS_2(g)] = 115.3$ kJ/mol $= [716.67 + 2 \times 276.98] - [2 \times \text{C=S}]$
The C=S double bond energy is therefore given by the equation:
C=S $= -(115.3 - 716.67 - 2 \times 276.98) \div 2 = 577.7$ kJ/mol

18.102   There are six S—F bonds in the molecule:

$\Delta H_f° = \text{sum}(\Delta H_f°[\text{gaseous atoms}]) - \text{sum}(\text{average bond energies in the molecule})$

$\Delta H_f°[SF_6(g)] = -1096$ kJ/mol $= [276.98 + 6 \times 79.14] - [6 \times \text{S—F}]$

S—F $= (1096 + 276.98 + 6 \times 79.14) \div 6 = 308.0$ kJ/mol

18.104   We must consider two S=O bonds and two S–F bonds:

ΔH_f° = sum(ΔH_f°[gaseous atoms]) – sum(average bond energies in the molecule)

$\Delta H_f^\circ[SO_2F_2(g)]$ = –858 kJ/mol = [274.7 + 2 × 249.2 + 2 × 78.91] –

[2 × 307.4 + 2 × S=O]

S=O = 587 kJ/mol

18.106 ΔH_f° = sum(ΔH_f°[gaseous atoms]) – sum(average bond energies in the molecule)

$\Delta H_f^\circ[CCl_4(g)]$ = [716.67 + 4 × 121.47] – [4 × 338] = –149 kJ/mol

Additional Exercises

18.113   No work (P × ΔV) is accomplished in a bomb calorimeter because there is no change in volume. At constant pressure, w = $-\Delta n_{gas}RT$ = –(18 mol – 15 mol)(8.314 J mol^{-1} K^{-1})(298 K) = –7.43 x 10^3 J = –7.43 kJ

18.114   We can calculate the work of the expanding gas (PΔV) if we can calculate the change in volume ΔV. Since the initial volume is given (5.00 L), we need only to calculate the final volume. For this, it is first necessary to determine the value of n, the number of moles of gas.

$$n = \frac{PV}{RT} = \frac{(4.00\,atm)(5.00\,L)}{\left(0.0821\,\frac{L\,atm}{mol\,K}\right)(298\,K)} = 0.817\,mol$$

$$V_2 = \frac{nRT}{P_2} = \frac{(0.817\,mol)\left(0.0821\,\frac{L\,atm}{mol\,K}\right)(298\,K)}{1\,atm} = 20.0\,L$$

$$\Delta V = (20.0\,L - 5.00\,L) = 15.0\,L$$

The work of gas expansion against a constant pressure of 1 atm is then given by the quantity –PΔV: w = –(1 atm)(15.0 L) = –15.0 L atm

18.117   First, review the information provided. Since the salt dissolved, the process is spontaneous, and the sign for ΔG is negative. The temperature went down indicating that this is an endothermic process and ΔH must be positive.. Dissolving the solid salt increases the disorder of the system which indicates ΔS is positive. The general equation from which we must work is $\Delta G = \Delta H - T\Delta S$. Using this equation, the magnitude of TΔS must be larger than the magnitude of ΔH in order to obtain a negative value for ΔG.

18.120   (a) $H_2C_2O_{4(s)}$ + 1/2$O_{2(g)}$ → $H_2O(l)$ + 2$CO_{2(g)}$ $\Delta H°_{comb}$ = –246.05 kJ

(b) $H_{2(g)}$ + 2$C_{(s)}$ + 2$O_{2(g)}$ → $H_2C_2O_{4(s)}$

(c) $\Delta H°_{comb}$ = {$\Delta H_f^\circ(H_2O_{(l)})$ + 2$\Delta H_f^\circ(CO_{2(g)})$} – {$\Delta H_f^\circ(H_2C_2O_{4(s)})$ + 1/2$\Delta H_f^\circ(O_{2(g)})$}
Rearranging;
$\Delta H_f^\circ(H_2C_2O_{4(s)})$ = {$\Delta H_f^\circ(H_2O_{(l)})$ + 2$\Delta H_f^\circ(CO_{2(g)})$}
$-\Delta H°$ comb
= –285.8 kJ + 2(–393.5 kJ) – (246.05 kJ) = –826.8 kJ

(d) ΔS_f° = $S°(H_2C_2O_{4(s)})$ = {$S_f°(H_{2(g)})$ + 2$S°(C_{(g)})$ + 2$S°(O_{2(g)})$}
= 120.1 J/K – {130.6J/K + 2(5.69 J/K) + 2(205.0 J/K)}
= –431.9J/mol K
$\Delta S°_{comb}$ = {$S°(H_2O_{(l)})$ + 2$S°(CO_{2(g)})$} – {$S°(H_2C_2O_{4(s)})$ + 1/2$S°(O_{2(g)})$}
= {69.96 J/K + 2(213.6 J/K)} – {120.1 J/K + ½(205.0 J/K) = 274.6 J/mol K

(e) $\Delta G_f^\circ = \Delta H_f^\circ - T\Delta S_f^\circ$
= –826.8 kJ – (298 K)(–431.9 J/K) = –698.1 kJ
$\Delta G_{comb}° = \Delta H_{comb}° - T\Delta S_{comb}°$
= –246.05 kJ –(298K)(274.6 J/K) = –544.3 kJ

18.123　We need to calculate the amount of energy produced when one gallon of each of these fuels is burned;

$$\text{\# mol ethanol} = \left(3.78 \times 10^3 \text{ mL ethanol}\right)\left(\frac{0.7893 \text{ g ethanol}}{1 \text{ mL ethanol}}\right)\left(\frac{1 \text{ mole ethanol}}{46.07 \text{ g ethanol}}\right)$$

$$= 64.8 \text{ moles ethanol}$$

$$\text{\# kJ} = (64.8 \text{ moles ethanol})\left(\frac{-1299.8 \text{ kJ}}{1 \text{ mole ethanol}}\right) = 8.42 \times 10^4 \text{ kJ}$$

$$\text{\# mol octane} = \left(3.78 \times 10^3 \text{ mL octane}\right)\left(\frac{0.7025 \text{ g octane}}{1 \text{ mL octane}}\right)\left(\frac{1 \text{ mole octane}}{114.23 \text{ g octane}}\right)$$

$$= 23.2 \text{ moles octane}$$

$$\text{\# kJ} = (23.2 \text{ moles octane})\left(\frac{-5307 \text{ kJ}}{1 \text{ mole octane}}\right) = 1.23 \times 10^5 \text{ kJ}$$

In spite of the large number of moles of ethanol in one gallon of liquid, the energy produced from the combustion of a gallon of octane is greater than the amount produced when one gallon of ethanol is burned.

18.125
$$\begin{array}{ccccc}
\text{C(s)} & + & 2\text{Cl}_2(g) & \rightarrow & \text{CCl}_4(\ell) \\
\downarrow(1) & & \downarrow(2) \quad (3) & & \uparrow(4) \\
\text{C(g)} & + & 4\text{Cl}(g) & \rightarrow & \text{CCl}_4(g)
\end{array}$$

Step 1:　$\Delta H = 716.67 \text{ kJ}$
Step 2:　$\Delta H = 4(121.5 \text{ kJ}) = 485.88 \text{ kJ}$
Step 3:　$\Delta H_f^\circ(\text{CCl}_4(g)) = 4(\text{BE(C–Cl)}) = 4(338 \text{ kJ}) = -1350 \text{ kJ}$
Step 4:　$\Delta H_{cond} = -29.9 \text{ kJ}$

ΔH = the sum of the four steps = -179 kJ

Practice Exercises

P19.1 $2H_2O(\ell) + 2e^- \rightarrow H_2(g) + 2OH^-(aq)$ reduction
       $2Br^-(aq) \rightarrow Br_2(aq) + 2e^-$ oxidation
       $2H_2O(\ell) + 2Br^-(aq) \rightarrow Br_2(aq) + H_2(g) + 2OH^-(aq)$ net cell reaction

P19.2 The number of Coulombs is: $4.00\ A \times 200\ s = 800\ C$
       The number of moles is:

$$\# \text{ mol } OH^- = 800\ C \times \frac{1\ F}{96,500\ C} \times \frac{1\ \text{mol } OH^-}{1\ F} = 8.29 \times 10^{-3}\ \text{mol } OH^-$$

P19.3 The number of moles of Au to be deposited is: $3.00\ g\ Au \div 197\ g/mol = 0.0152\ mol\ Au$. The number of Coulombs ($A\bullet s$) is:

$$\# \ C = 0.0152\ \text{mol Au} \times \frac{3\ F}{1\ \text{mol Au}} \times \frac{96,500\ C}{1\ F} = 4.40 \times 10^3\ C$$

The number of minutes is:

$$\# \min = \frac{4.40 \times 10^3\ A \cdot s}{10.0\ A} \times \frac{1\ \min}{60\ s} = 7.33\ \min$$

P19.4 As in Practice Exercise 3 above, the number of Coulombs is $4.40 \times 10^3\ C$. This corresponds to a current of:

$$\# \ A = \frac{4.40 \times 10^3\ A \cdot s}{20.0\ \min} \times \frac{1\ \min}{60\ s} = 3.67\ A$$

P19.5 anode: $Mg(s) \rightarrow Mg^{2+}(aq) + 2e^-$
       cathode: $Fe^{2+}(aq) + 2e^- \rightarrow Fe(s)$
       cell notation: $Mg(s) \,|\, Mg^{2+}(aq) \,||\, Fe^{2+}(aq) \,|\, Fe(s)$

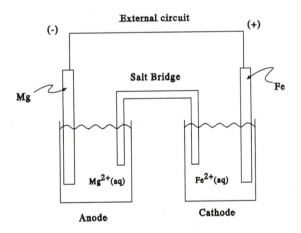

P19.6 anode: $Al(s) \rightarrow Al^{3+}(aq) + 3e^-$
       cathode: $Pb^{2+}(aq) + 2e^- \rightarrow Pb(S)$

P19.7 $E°_{cell} = E°_{substance\ reduced} - E°_{substance\ oxidized}$

$1.93\ V = (-0.44\ V) - E°_{Mg}2+$

$E°_{Mg}2+ = -0.44 - 1.93 = -2.37\ V$

This agrees exactly with Table 19.1.

P19.8   The half–reaction with the more positive value of E° (listed higher in Table 19.1) will occur as a reduction. The half–reaction having the less positive (more negative) value of E° (listed lower in Table 19.1) will be reversed and occur as an oxidation.

$Br_2(aq) + 2e^- \rightarrow 2Br^-(aq)$	reduction
$H_2SO_3(aq) + H_2O(\ell) \rightarrow SO_4^{2-}(aq) + 4H^+(aq) + 2e^-$	oxidation
$Br_2(aq) + H_2SO_3(aq) + H_2O(\ell) \rightarrow 2Br^-(aq) + SO_4^{2-}(aq) + 4H^+(aq)$	

P19.9   The half–reaction having the more positive value for E° will occur as a reduction. The other half–reaction should be reversed, so as to appear as an oxidation.

$NiO_2(s) + 2H_2O(\ell) + 2e^- \rightarrow Ni(OH)_2(s) + 2OH^-(aq)$	reduction
$Fe(s) + 2OH^-(aq) \rightarrow 2e^- + Fe(OH)_2(s)$	oxidation
$NiO_2(s) + Fe(s) + 2H_2O(\ell) \rightarrow Ni(OH)_2(s) + Fe(OH)_2(s)$	net reaction

$E°_{cell} = E°_{substance\ reduced} - E°_{substance\ oxidized}$

$E°_{cell} = E°_{NiO}2 - E°_{Fe}$

$E°_{cell} = 0.49 - (-0.88) = 1.37\ V$

P19.10   The half–reaction having the more positive value for E° will occur as a reduction. The other half–reaction should be reversed, so as to appear as an oxidation.

$3 \times [MnO_4^-(aq) + 8H^+(aq) + 5e^- \rightarrow Mn^{2+}(aq) + 4H_2O(\ell)]$	reduction
$5\ [Cr(s) \rightarrow Cr^{3+}(aq) + 3e^-]$	oxidation
$3MnO_4^-(aq) + 24H^+(aq) + 5Cr(s) \rightarrow 5Cr^{3+}(aq) + 3Mn^{2+}(aq) + 12H_2O(\ell)$	net reaction

$E°_{cell} = E°_{substance\ reduced} - E°_{substance\ oxidized}$

$E°_{cell} = E°_{MnO}4^- - E°_{Cr}$

$E°_{cell} = 1.51\ V - (-0.74\ V) = 2.25\ V$

P19.11   A reaction will occur spontaneously in the forward direction if the value of E° is positive. We therefore evaluate E° for each reaction using:

$E°_{cell} = E°_{substance\ reduced} - E°_{substance\ oxidized}$

(a) $Br_2(aq) + 2e^- \rightarrow 2Br^-(aq)$ reduction

        $Cl_2(aq) + 2H_2O(\ell) \rightarrow 2HOCl(aq) + 2H^+(aq) + 2e^-$ oxidation

        $E°_{cell} = E°_{Br}2 - E°_{Cl}2$

        $E°_{cell} = 1.07\ V - (1.36\ V) = -0.29\ V$ $\therefore$ nonspontaneous

(b) $2Cr^{3+}(aq) + 6e^- \rightarrow 2Cr(s)$ reduction

        $3Zn(s) \rightarrow 3Zn^{2+}(aq) + 6e^-$ oxidation

        $E°_{cell} = E°_{Cr}3+ - E°_{Zn}$

$$E°_{cell} = -0.74 \text{ V} - (-0.76 \text{ V}) = +0.02 \text{ V} \quad \text{spontaneous}$$

P19.12   The possible reactions are:

$$Cd^{2+} + 2e^- \rightleftharpoons Cd_{(s)} \qquad\qquad E° = -0.40 \text{ V}$$
$$Sn^{2+} + 2e^- \rightleftharpoons Sn_{(s)} \qquad\qquad E° = -0.14 \text{ V}$$
$$2H_2O + 2e^- \rightleftharpoons H_{2(g)} + 2OH^-_{(aq)} \qquad E° = -0.83 \text{ V}$$

Choose the one with the least negative value.

$$Sn^{2+} + 2e^- \rightleftharpoons Sn_{(s)}$$

P19.13   It was stated in Practice Exercise 7 that the reaction in question had a standard cell potential of 1.96 V, or 1.96 J/C. Since 2 mole of e^- are involved, i.e., n = 2, we have:

$$G° = -nFE°_{cell} = -(2)(96{,}500 \text{ C})(1.96 \text{ V}) = -3.78 \text{ X } 10^5 \text{ J} = -3.78 \text{ X } 10^2 \text{ kJ}$$

P19.14   Using Equation 19.7

$$E°_{cell} = \frac{0.0592 \text{ V}}{n} \log K_c$$

$$-0.46 \text{ V} = \frac{0.0592 \text{ V}}{2} \log K_c$$

$$\log K_c = -15.5$$

Taking the antilog of both sides of the above equation gives:
$$K_c = 2.9 \text{ X } 10^{-16}$$

This very small value for the equilibrium constant means that the products of the reaction are not formed spontaneously. The equilibrium lies far to the left, favoring reactants, and we do not expect much product to form.

P19.15
$$Zn(s) \; Zn^{2+}(aq) + 2e^- \qquad \text{oxidation}$$
$$Cu^{2+}(aq) + 2e^- \rightarrow 2Cu(s) \qquad \text{reduction}$$

$$E°_{cell} = E°_{Cu}2+ - E°_{Zn}$$
$$E°_{cell} = +0.34 \text{ V} - (-0.76 \text{ V}) = +1.10 \text{ V}$$

The Nernst equation for this cell is:

$$E_{cell} = E°_{cell} - \frac{0.0592 \text{ V}}{n} \log \frac{\left|Zn^{2+}\right|}{\left|Cu^{2+}\right|}$$

$$= 1.10 \text{ V} - \frac{0.0592 \text{ V}}{2} \log \frac{[1.0]}{[0.010]}$$

$$= 1.04 \text{ V}$$

P19.16
$$Cu(s) \rightarrow Cu^{2+}(aq) + 2e^- \qquad \text{oxidation}$$
$$Ag^+(aq) + e^- \rightarrow Ag(s) \qquad\qquad \text{reduction}$$

$$E°_{cell} = E°_{Ag}+ - E°_{Cu}$$
$$E°_{cell} = +0.80 \text{ V} - (+0.34 \text{ V}) = +0.46 \text{ V}$$

$$E_{cell} = E^\circ_{cell} - \frac{0.0592\,V}{n} \log \frac{\left[Cu^{2+}\right]}{\left[Ag^+\right]^2}$$

$$\log \frac{\left[Cu^{2+}\right]}{\left[Ag^+\right]^2} = \frac{\left(E^\circ_{cell} - E_{cell}\right)}{\frac{0.0592\,V}{n}}$$

$$= \frac{\left(0.46\,V - 0.57\,V\right)}{\frac{0.0592\,V}{2}}$$

$$= -3.71$$

$$\frac{\left[Cu^{2+}\right]}{\left[Ag^+\right]^2} = 10^{-3.71} = 1.9 \times 10^{-4}$$

Since the $[Ag^+] = 1.00$ M, $[Cu^{2+}] = 1.9 \times 10^{-4}$ M

Substituting the second value into the same expression gives $[Cu^{2+}] = 6.9 \times 10^{-13}$ M

P19.17   We are told that, in this galvanic cell, the chromium electrode is the anode, meaning that oxidation occurs at the chromium electrode.

Now in general, we have the equation:
$$E^\circ_{cell} = E^\circ_{reduction} - E^\circ_{oxidation}$$
which becomes, in particular for this case:
$$E^\circ_{cell} = E^\circ_{Ni^{2+}} - E^\circ_{Cr}$$

The net cell reaction is given by the sum of the reduction and the oxidation half–reactions, multiplied in each case so as to eliminate electrons from the result:

3 $[Ni^{2+}(aq) + 2e^- \rightarrow Ni(s)]$	reduction
2 $[Cr(s) \rightarrow Cr^{3+}(aq) + 3e^-]$	oxidation
$3Ni^{2+}(aq) + 2Cr(s) \rightarrow 2Cr^{3+}(aq) + 3Ni(s)$	net reaction

In this reaction, n = 6, and the Nernst equation becomes:

$$E_{cell} = E^\circ_{cell} - \frac{0.0592\,V}{n} \log \frac{\left[Cr^{3+}\right]^2}{\left[Ni^{2+}\right]^3}$$

$$\log \frac{\left[Cr^{3+}\right]^2}{\left[Ni^{2+}\right]^3} = \frac{\left(E^\circ_{cell} - E_{cell}\right)}{\frac{0.0592\,V}{n}}$$

$$= \frac{\left(0.487\,V - 0.552\,V\right)}{\frac{0.0592\,V}{6}}$$

$$= -6.59$$

$$\frac{\left[Cr^{3+}\right]^2}{\left[Ni^{2+}\right]^3} = 10^{-6.59} = 2.6 \times 10^{-7}$$

Substituting $[Ni^{2+}] = 1.20$ M, we solve for $[Cr^{3+}]$ and get: $[Cr^{3+}] = 6.7 \times 10^{-4}$ M.

Review Problems

19.54   1 C = 1 As
   (a)   4.00 A 600 s = 2.40 X10³ C
   (b)   10.0 A 20 min 60 s/min = 1.2 X10⁴ C
   (c)   1.50 A 6.00 hr 3600 s/hr = 3.24 X10⁴ C

19.56   (a) $Fe^{2+}(aq) + 2e^- \rightarrow Fe(s)$
               0.20 mol Fe^{2+} 2 mol e⁻/mol Fe^{2+} = 0.40 mol e⁻
         (b) $Cl^-(aq) \rightarrow Cl_2(g) + e^-$
               0.70 mol Cl⁻ 1 mol e⁻/mol Cl⁻ = 0.70 mol e⁻
         (c) $Cr^{3+}(aq) + 3e^- \rightarrow Cr(s)$
               1.50 mol Cr^{3+} 3 mol e⁻/mol Cr^{3+} = 4.50 mol e⁻
         (d) $Mn^{2+}(aq) + 4H_2O(\ell) \rightarrow MnO_4^-(aq) + 8H^+(aq) + 5e^-$
               1.0 X10⁻² mol Mn^{2+} 5 mol e⁻/mol Mn^{2+} = 5.0 X10⁻² mol e⁻

19.58 $Ag^+(aq) + e^- \rightarrow Ag(s)$, and $Cr^{3+}(aq) + 3e^- \rightarrow Cr(s)$
   This shows that there are three moles of electrons per mole of Cr but only one mole of electrons per mole of
   Ag. The number of moles of electrons involved in the silver reaction is:

$$\# \, mol \, e^- = (12.0 \, g \, Ag)\left(\frac{1 \, mol \, Ag}{107.9 \, g \, Ag}\right)\left(\frac{1 \, mol \, e^-}{1 \, mol \, Ag}\right) = 0.111 \, mol \, e^-$$

   The amount of Cr is then:

$$\# \, mol \, Cr^{3+} = (0.111 \, mol \, e^-)\left(\frac{1 \, mol \, Cr^{3+}}{3 \, mol \, e^-}\right) = 0.0371 \, mol \, Cr^{3+}$$

19.60 $Fe(s) + 2OH^-(aq) \rightarrow Fe(OH)_2(s) + 2e^-$
   The number of Coulombs is: 12.0 min 60 s/min 8.00 C/s = 5.76 X10³ C. The number of grams of $Fe(OH)_2$
   is:

$$\# \, g \, Fe(OH)_2 = (5.76 \, X \, 10^3 \, C)\left(\frac{1 \, mol \, e^-}{96500 \, C}\right)\left(\frac{1 \, mol \, Fe(OH)_2}{2 \, mol \, e^-}\right)\left(\frac{89.86 \, g \, Fe(OH)_2}{1 \, mol \, Fe(OH)_2}\right)$$

$$= 2.68 \, g \, Fe(OH)_2$$

19.62   19.26 $Cr^{3+}(aq) + 3e^- \rightarrow Cr(s)$
   The number of Coulombs that will be required is:

$$\# \, C = (75.0 \, g \, Cr)\left(\frac{1 \, mol \, Cr}{52.00 \, g \, Cr}\right)\left(\frac{3 \, mol \, e^-}{1 \, mol \, Cr}\right)\left(\frac{96500 \, C}{1 \, mol \, e^-}\right) = 4.18 \, X \, 10^5 \, C$$

   The time that will be required is:

$$\# \, hr = (4.18 \, X \, 10^5 \, C)\left(\frac{1 \, s}{2.25 \, C}\right)\left(\frac{1 \, hr}{3600 \, s}\right) = 51.5 \, hr$$

19.64 $Mg^{2+}(aq) + 2e^- \rightarrow Mg(\ell)$
The number of Coulombs that will be required is:

$$\# C = (60.0 \text{ g Mg})\left(\frac{1 \text{ mol Mg}}{24.31 \text{ g Mg}}\right)\left(\frac{2 \text{ mol e}^-}{1 \text{ mol Mg}}\right)\left(\frac{96500 \text{ C}}{1 \text{ mol e}^-}\right) = 4.76 \times 10^5 \text{ C}$$

The number of amperes is: 4.76×10^5 C 7200 s = 66.2 amp

19.66   The electrolysis of NaCl solution results in the reduction of water, together with the formation of hydroxide ion: $2H_2O(\ell) + 2e^- \rightarrow H_2(g) + 2OH^-(aq)$. The number of Coulombs is: 2.50 A 15.0 min 60 s/min = 2.25 $\times 10^3$ C. The number of moles of OH$^-$ is:

$$\# \text{ mol OH}^- = (2.25 \times 10^3 \text{ C})\left(\frac{1 \text{ mol e}^-}{96500 \text{ C}}\right)\left(\frac{2 \text{ mol OH}^-}{2 \text{ mol e}^-}\right) = 0.0233 \text{ mol OH}^-$$

The volume of acid solution that will neutralize this much OH$^-$ is:

$$\# \text{ mL HCl} = (0.0233 \text{ mol OH}^-)\left(\frac{1 \text{ mol HCl}}{1 \text{ mol OH}^-}\right)\left(\frac{1000 \text{ mL HCl}}{0.100 \text{ mol HCl}}\right) = 233 \text{ mL HCl}$$

19.68   (a)   anode: $Cd(s) \rightarrow Cd^{2+}(aq) + 2e^-$
              cathode: $Au^{3+}(aq) + 3e^- \rightarrow Au(s)$
              cell: $3Cd(s) + 2Au^{3+}(aq) \rightarrow 3Cd^{2+}(aq) + 2Au(s)$

       (b)   anode: $Pb(s) + SO_4^{2-}(aq) \rightarrow PbSO_4(s) + 2e^-$
              cathode: $PbO_2(s) + SO_4^{2-}(aq) + 4H^+(aq) + 2e^- \rightarrow PbSO_4(s) + 2H_2O(\ell)$
              cell: $Pb(s) + PbO_2(s) + 2SO_4^{2-}(aq) + 4H^+(aq) \rightarrow 2PbSO_4(s) + 2H_2O(\ell)$

       (c)   anode: $Cr(s) \rightarrow Cr^{3+}(aq) + 3e^-$
              cathode: $Cu^{2+}(aq) + 2e^- \rightarrow Cu(s)$
              cell: $2Cr(s) + 3Cu^{2+}(aq) \rightarrow 2Cr^{3+}(aq) + 3Cu(s)$

19.70   (a) $Fe(s)Fe^{2+}(aq)Cd^{2+}(aq)Cd(s)$
       (b) $Pt(s),Cl^-(aq)Cl_2(g)Br_2(aq)Br^-(aq), Pt(s)$
       (c) $Ag(s)Ag^+(aq)Au^{3+}(aq)Au(s)$

19.72   (a) $Sn(s)$ (b) $Br^-(aq)$ (c) $Zn(s)$ (d) $I^-(aq)$

19.76   The reactions are spontaneous if the overall cell potential is positive.
$E^\circ_{cell} = E^\circ_{substance\ reduced} - E^\circ_{substance\ oxidized}$

       (a) $E^\circ_{cell} = 1.42 \text{ V} - (0.54 \text{ V}) = 0.88 \text{ V}$ spontaneous
       (b) $E^\circ_{cell} = -0.44 \text{ V} - (0.96 \text{ V}) = -1.40 \text{ V}$ not spontaneous
       (c) $E^\circ_{cell} = -0.74 \text{ V} - (-2.76 \text{ V}) = 2.02 \text{ V}$ spontaneous

19.78   The given equation is separated into its two half reactions:
$MnO_4^-(aq) + 8H^+(aq) + 5e^- \rightarrow Mn^{2+}(aq) + 4H_2O(\ell)$ reduction
$5Fe^{2+}(aq) \rightarrow 5Fe^{3+}(aq) + 5e^-$ oxidation

$E^\circ_{cell} = E^\circ_{reduction} - E^\circ_{oxidation} = 1.51 \text{ V} - 0.77 \text{ V} = 0.74 \text{ V}$

19.80   (a) $Zn(s)Zn^{2+}(aq)Co^{2+}(aq)Co(s)$

                 $E^{\circ}_{cell} = -0.28\ V - (-0.76\ V) = 0.48\ V$

  (b) $Mg(s)Mg^{2+}(aq)Ni^{2+}(aq)Ni(s)$

                 $E^{\circ}_{cell} = -0.25\ V - (-2.37\ V) = 2.12\ V$

  (c) $Sn(s)Sn^{2+}(aq)Au^{3+}(aq)Au(s)$

                 $E^{\circ}_{cell} = 1.42\ V - (-0.14\ V) = 1.56\ V$

19.82   The half cell with the more positive E°_{cell} will appear as a reduction, and the other half reaction is reversed, to appear as an oxidation:

     $BrO_3^-(aq) + 6H^+(aq) + 6e^- \rightarrow Br^-(aq) + 3H_2O$ reduction

     $3\ (2I^-(aq) \rightarrow I_2(s) + 2e^-)$ oxidation

     $BrO_3^-(aq) + 6I^-(aq) + 6H^+(aq) \rightarrow 3I_2(s) + Br^-(aq) + 3H_2O$ net reaction

     $E^{\circ}_{cell} = E^{\circ}_{substance\ reduced} - E^{\circ}_{substance\ oxidized}$ or

     $E^{\circ}_{cell} = E^{\circ}_{reduction} - E^{\circ}_{oxidation} = 1.44\ V - (0.54\ V) = 0.90\ V$

19.84   The half reaction having the more positive standard reduction potential is the one that occurs as a reduction, and the other one is written as an oxidation:

     $2\ (2HOCl(aq) + 2H^+(aq) + 2e^- \rightarrow Cl_2(g) + 2H_2O(\ell))$ reduction

     $3H_2O(\ell) + S_2O_3^{2-}(aq) \rightarrow 2H_2SO_3(aq) + 2H^+(aq) + 4e^-$ oxidation

     $4HOCl(aq) + 4H^+(aq) + 3H_2O(\ell) + S_2O_3^{2-}(aq) \rightarrow$

                               $2Cl_2(g) + 4H_2O(\ell) + 2H_2SO_3(aq) + 2H^+(aq)$

   which simplifies to give the following net reaction:

     $4HOCl(aq) + 2H^+(aq) + S_2O_3^{2-}(aq) \rightarrow 2Cl_2(g) + H_2O(\ell) + 2H_2SO_3(aq)$

19.86   The two half reactions are:

     $SO_4^{2-}(aq) + 2e^- + 4H^+(aq) \rightarrow H_2SO_3(aq) + H_2O(\ell)$ reduction

     $2I^-(aq) \rightarrow I_2(s) + 2e^-$ oxidation

     $E^{\circ}_{cell} = E^{\circ}_{reduction} - E^{\circ}_{oxidation} = 0.17\ V - (0.54\ V) = -0.37\ V$

   Since the overall cell potential is negative, we conclude that the reaction is not spontaneous in the direction written.

19.88   Possible cathode reactions:

     $Al^{3+} + 3e^- \rightleftharpoons Al_{(s)}$ $E^{\circ} = -1.66\ V$

     $2H_2O + 2e^- \rightleftharpoons H_{2(g)} + 2OH^-_{(aq)}$ $E^{\circ} = -0.83\ V$

   Possible anode reactions:

     $S_2O_8^{2-} + 2e^- \rightleftharpoons 2SO_4^{2-}$ $E^{\circ} = +2.05\ V$

     $O_2 + 4H^+ + 4e^- \rightleftharpoons 2H_2O$ $E^{\circ} = +1.23\ V$

   Cathode reaction:

     $2H_2O + 2e^- \rightleftharpoons H_{2(g)} + 2OH^-_{(aq)}$ $E^{\circ} = -0.83\ V$

   Anode reactions:

     $2H_2O \rightleftharpoons O_2 + 4H^+ + 4e^-$ $E^{\circ} = -1.23\ V$

     $2H_2O \rightleftharpoons 2H_{2(g)} + O_{2(g)}$ $E^{\circ} = -2.06\ V$

19.90 First, separate the overall reaction into its two half reactions:

$2Br^-(aq) \rightarrow Br_2(aq) + 2e^-$ oxidation

$I_2(s) + 2e^- \rightarrow 2I^-(aq)$ reduction

$E^\circ_{cell} = E^\circ_{reduction} - E^\circ_{oxidation} = 0.54\ V - (1.07\ V) = -0.53\ V$

The value of n is 2: $G^\circ = -nFE^\circ_{cell} = -(2)(96,500\ C)(-0.53\ J/C)$
$= 1.0 \times 10^5\ J = 1.0 \times 10^2\ kJ$

19.92 (a) $E^\circ_{cell} = E^\circ_{reduction} - E^\circ_{oxidation} = 2.01\ V - (1.47\ V) = 0.54\ V$

(b) Since n = 10, $G^\circ = -nFE^\circ_{cell} = -(10)(96,500\ C)(0.54\ J/C) = -5.2 \times 10^5\ J$
$G^\circ = -5.2 \times 10^2\ kJ$

(c) $E^\circ_{cell} = \dfrac{0.0592}{n} \log K_c$

$0.54\ V = (0.0592\ V)/10\ \log K_c$
$\log K_c = 91.22$ Taking the antilog of both sides of this equation:
$K_c = 1.7 \times 10^{91}$

19.94 Sn is oxidized by two electrons and Ag is reduced by two electrons:

$E^\circ_{cell} = \dfrac{0.0592}{n} \log K_c$
$-0.015\ V = 0.0592\ V/2\ \log K_c$
$\log K_c = -0.51$
$K_c = antilog(-0.51) = 0.31$

19.96 This reaction involves the oxidation of Ag by two electrons and the reduction of Ni by two electrons. The concentration of the hydrogen ion is derived from the pH of the solution: $[H^+] = antilog(-pH) = antilog(-5) = 1 \times 10^{-5}\ M$

$E_{cell} = 2.48\ V - \dfrac{0.0592\ V}{2} \log \dfrac{[Ag^+]^2[Ni^{2+}]}{[H^+]^4}$

$= 2.48\ V - \dfrac{0.0592\ V}{2} \log \dfrac{[1.0 \times 10^{-2}]^2[1.0 \times 10^{-2}]}{[1.0 \times 10^{-5}]^4}$

$E_{cell} = 2.48\ V - 0.41\ V = 2.07\ V$

19.98

$E_{cell} = E^\circ_{cell} - \dfrac{0.0592\ V}{n} \log Q$

$E_{cell} = E^\circ_{cell} - \dfrac{0.0592\ V}{2} \log \dfrac{[Mg^{2+}]}{[Cd^{2+}]}$

$1.54\ V = 1.97\ V - \dfrac{0.0592\ V}{2} \log \dfrac{1}{[Cd^{2+}]}$

$\log (1/[Cd^{2+}]) = 14.53$, and $[Cd^{2+}] = 3.0 \times 10^{-15}\ M$

19.100 In the iron half cell, we are initially given:
0.0500 mL 0.100 mol/L = 5.00×10^{-3} mol Fe^{2+}(aq)

The precipitation of $Fe(OH)_2$(s) consumes some of the added hydroxide ion, as well as some of the iron ion: Fe^{2+}(aq) + 2OH⁻(aq) → $Fe(OH)_2$(s). The number of moles of OH⁻ that have been added to the iron half cell is:
0.500 mol/L 0.0500 L = 2.50×10^{-2} mol OH⁻

The stoichiometry of the precipitation reaction requires that the following number of moles of OH⁻ be consumed on precipitation of 5.00×10^{-3} mol of $Fe(OH)_2$(s):
5.00×10^{-3} mol $Fe(OH)_2$ 2 mol OH⁻/mol $Fe(OH)_2$ = 1.00×10^{-2} mol OH⁻

The number of moles of OH⁻ that are unprecipitated in the iron half cell is:
2.50×10^{-2} mol − 1.00×10^{-2} mol = 1.50×10^{-2} mol OH⁻

Since the resulting volume is 50.0 mL + 50.0 mL, the concentration of hydroxide ion in the iron half cell becomes, upon precipitation of the $Fe(OH)_2$:
[OH⁻] = 1.50×10^{-2} mol/0.100 L = 0.150 M OH⁻

We have assumed that the iron hydroxide that forms in the above precipitation reaction is completely insoluble. This is not accurate, though, because some small amount does dissolve in water according to the following equilibrium:
$Fe(OH)_2$(s) → Fe^{2+}(aq) + 2OH⁻(aq)

This means that the true [OH⁻] is slightly higher than 0.150 M as calculated above. Thus we must set up the usual equilibrium table, in order to analyze the extent to which $Fe(OH)_2$(s) dissolves in 0.150 M OH⁻ solution:

	$[Fe^{2+}]$	$[OH^-]$
I	–	0.150
C	+x	+2x
E	+x	0.150+2x

The quantity x in the above table is the molar solubility of $Fe(OH)_2$ in the solution that is formed in the iron half cell.

$K_{sp} = [Fe^{2+}][OH^-]^2 = (x)(0.150 + 2x)^2$

The standard cell potential is:
$E°_{cell} = E°_{reduction} − E°_{oxidation} = 0.3419 − (−0.447) = 0.7889$

The Nernst equation is:

$$E_{cell} = E°_{cell} − \frac{0.0592\,V}{n} \log \frac{\left[Fe^{2+}\right]}{\left[Cu^{2+}\right]}$$

$$1.175\,V = 0.7889\,V − \frac{0.0592\,V}{2} \log \frac{\left[Fe^{2+}\right]}{1.00\,M}$$

which becomes: $1.175 = 0.7889 − 0.0296 \log [Fe^{2+}]$
$[Fe^{2+}] = 9.04 \times 10^{-14}$ M

This is the concentration of Fe^{2+} in the saturated solution, and it is the value to be used for x in the above expression for K_{sp}.

$K_{sp} = (x)(0.150 + 2x)^2 = (9.04 \times 10^{-14})[0.150 + (2)(9.04 \times 10^{-14})]^2$
$K_{sp} = 2.03 \times 10^{-15}$

Additional Exercises

19.103   (a)   First, we calculate the number of Coulombs:
                1.50 A 30.0 min 60 s/min = 2.70×10^3 As = 2.70×10^3 C
                Then we determine the number of moles of electrons:

$$\text{\# mol } e^- = \left(2.70 \times 10^3 \text{ C}\right)\left(\frac{1 \text{ mol } e^-}{96500 \text{ C}}\right) = 0.0280 \text{ mol } e^-$$

  (b)   0.475 g 50.9 g/mol = 9.33×10^{-3} mol V
  (c) $(2.80 \times 10^{-2}$ mol $e^-)/(9.33 \times 10^{-3}$ mol V) = 3.00 mol e^-/mol V
            The original oxidation state was V^{3+}.

19.106   The half reactions are:
                $2Cl^- \rightarrow Cl_2 + 2e^-$
                $2H_2O + 2e^- \rightarrow H_2 + 2OH^-$

The resulting pH (9.00) indicates a pOH of 5.00. This means that the concentration of hydroxide is: $[OH^-] = 1.00 \times 10^{-5}$ M

1.00×10^{-5} M 0.500 L = 5.00×10^{-6} moles OH^-
5.00×10^{-6} moles OH^- 2 mol e^-/2 mol OH^- = 5.00×10^{-6} mol e^-
5.00×10^{-6} mol e^- 96500 C/mol = 0.482 C
0.482 C 0.500 C/s = 0.964 s

19.107 $G° = -nFE°_{cell}$, $E°_{cell} = 1.34$ V = 1.34 J/C, and n = 2
        $G° = -(2)(96,500$ C)(1.34 J/C) = -2.59×10^5 J per mol of HgO

The maximum amount of work that can be derived from this cell, on using 1.00 g of HgO, is thus:

$$\text{\# J} = \left(1.00 \text{ g HgO}\right)\left(\frac{1 \text{ mol HgO}}{216.6 \text{ g HgO}}\right)\left(\frac{2.59 \times 10^5 \text{ J}}{1 \text{ mol HgO}}\right) = 1.20 \times 10^3 \text{ J}$$

Now, since 1 watt = 1 J s^{-1}, then 5×10^{-4} watt = 5×10^{-4} J s^{-1}, and the time required for this process is:

$$\text{\# hr} = \left(1.20 \times 10^3 \text{ J}\right)\left(\frac{1 \text{ s}}{5 \times 10^{-4} \text{ J}}\right)\left(\frac{1 \text{ min}}{60 \text{ s}}\right)\left(\frac{1 \text{ hr}}{60 \text{ min}}\right) = 6.67 \times 10^2 \text{ hr}$$

19.108   The initial numbers of moles of Ag^+ and Zn^{2+} are: 1.00 mol/L 0.100 L = 0.100 mol. The number of Coulombs (As) that have been employed is: 0.10 C/s 15.00 hr 3600 s/hr = 5.4×10^3 C. The number of moles of electrons is: 5.4×10^3 C 96,500 C/mol = 5.6×10^{-2} mol electrons.

For Ag^+, there is 1 mol per mole of electrons, and for Zn^{2+}, there are two moles of electrons per mol of Zn. This means that the number of moles of the two ions that have been consumed or formed is given by: 5.6×10^{-2} mol e^- 1 mol Ag^+/1 mol e^- = 5.6×10^{-2} mol Ag^+ reacted.

5.6×10^{-2} mol e^- 1 mol Zn^{2+}/2 mol e^- = 2.8×10^{-2} mol Zn^{2+} formed

The number of moles of Ag^+ that remain is: 0.100 − 0.056 = 0.044 mol of Ag^+

The final concentration of silver ion is: $[Ag^+]$ = 0.044 mol/0.100 L = 0.44 M

The number of moles of Zn^{2+} that are present is: 0.100 + 0.028 = 0.128 mol Zn^{2+}

The final concentration of zinc ion is: $[Zn^{2+}] = 0.128$ mol/0.100 L = 1.28 M

The standard cell potential should be: $E°_{cell} = E°_{reduction} - E°_{oxidation} = 0.80 - (-0.76) = 1.56$ V

We now apply the Nernst equation:

$$E_{cell} = E°_{cell} - \frac{0.0592\,V}{2} \log \frac{1.28}{(0.44)^2}$$

$E_{cell} = 1.56$ V $- 0.024$ V $= 1.54$ V

Practice Exercises

P20.1 $\Delta H = \Delta H_f^\circ = 601.7$ kJ

$\Delta S^\circ = S^\circ(Mg_{(s)}) + 1/2S^\circ(O_{2(g)}) - S^\circ(MgO_{(s)})$

     $= 32.5J/K + \frac{1}{2}(205 \ J/K) - 26.9 \ J/K$

     $= 108.1 \ J/K$

     $= 0.108 \ kJ/K$

$\Delta G_T^\circ = \Delta H^\circ - T\Delta S^\circ$

     $= 601.7 \ kJ - T(0.108 \ kJ/K)$

Decomposition occurs when $\Delta G < 0$. Solve for $\Delta G^\circ = 0$

$$T = \frac{601.7 \ kJ}{0.108 kJ/K} = 5570 \ K$$

P20.2   The net charge on the complex ion must first be determined. Two $S_2O_3^{2-}$ ions contribute a charge of 4–; the metal contributes a charge of 1+. The sum of these is 3–. The formula of the complex ion is therefore $[Ag(S_2O_3)_2]^{3-}$. The ammonium salt of this ion would have the formula $(NH_4)_3[Ag(S_2O_3)_2]$.

P20.3   The salt must include the six hydrated water molecules. We know that Al exists as a 3+ ion and that chloride has a charge of 1–. The hydrate would have the formula $AlCl_3 \bullet 6H_2O$. The complex ion most likely has the formula $[Al(H_2O)_6]^{3+}$.

P20.4   (a)   potassium hexacyanoferrate(III)

        (b)   dichlorobis(ethylenediamine)chromium(III) sulfate

P20.5   (a) $[SnCl_6]^{2-}$

        (b) $(NH_4)_2[Fe(CN)_4(H_2O)_2]$

P20.6   (a)   Since there are three ligands and $C_2O_4^{2-}$ is a bidentate ligand, the coordination number is 6.

        (b)   The coordination number is six. There are two bidentate ligands and two unidentate ligands.

        (c)   The coordination number is six. Both $C_2O_4^{2-}$ and ethylenediamine are bidentate ligands. Since there are three bidentate ligands, the coordination number must be six.

        (d)   EDTA is a hexadentate ligand so the coordination number is six.

Review Problems

20.64   First we need ΔH° for this reaction.

$\Delta H^\circ = 2\Delta H_f^\circ(Hg_{(g)}) + \Delta H_f^\circ(O_{2(g)}) - 2\Delta H_f^\circ(HgO_{(s)})$

     $= 2(61.3 \ kJ) + 0 - 2(-90.8 \ kJ) = 304 \ kJ$

Similarly;

$\Delta S^\circ = 2(175 \ J/K) + 205 \ J/K - (70.3 \ J/K) = 414 \ J/K$

Decomposition occurs when $\Delta G = 0$

$\Delta G^\circ = \Delta H^\circ - T\Delta S^\circ$

$$T = \frac{\Delta H^\circ}{\Delta S^\circ} = \frac{304 \ kJ}{0.414 \ kJ/K} = 734 \ K$$

20.65 $\Delta G = 0$ when $K_p = 1$

$\Delta H = -\Delta H_f^\circ(CuO) = 155$ kJ

$\Delta S = S^\circ(Cu_{(s)}) + 1/2S^\circ(O_{2(g)}) - S^\circ(CuO_{(s)})$

     $= 33.2 \ J/K + 1/2(205 \ J/K) - 42.6 \ J/K = 93.1 \ J/K$

$$\Delta G = \Delta H - T\Delta S = 0$$

$$T = \frac{\Delta H^\circ}{\Delta S^\circ} = \frac{155\,kJ}{0.0931\,kJ/K} = 1660\,K$$

20.66   (a) Bi_2O_5
       (b) PbS
       (c) PbI_2

20.68   (a) SnO
       (b) SnS

20.70   (a) $CaCl_2$
       (b) BeF_2

20.72   (a) HgS
       (b) Ag_2S

20.74   (a) HgS
       (b) Ag_2S

20.76   The net charge is 3^-, and the formula is $[Fe(CN)_6]^{3-}$.

20.78 $[CoCl_2(en)_2]^+$

20.80   (a)   oxalato   (b)   sulfido
       (c)   chloro   (d)   dimethylamine

20.82   (a)   hexaamminenickel(II) ion
       (b)   triamminetrichlorochromate(II) ion
       (c)   hexanitrocobaltate(III) ion
       (d)   diamminetetracyanomanganate(II) ion
       (e)   trioxalatoferrate(III) ion or trisoxalatoferrate(III) ion

20.84   (a) $[Fe(CN)_2(H_2O)_4]^+$
       (b) $Ni(C_2O_4)(NH_3)_4$
       (c) $[Fe(CN)_4(H_2O)_2]^-$
       (d) $K_3[Mn(SCN)_6]$
       (e) $[CuCl_4]^{2-}$

20.86   The coordination number is six, and the oxidation number of the iron atom is +2.

20.88

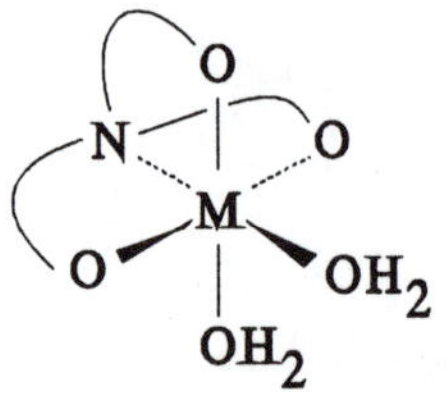

(The curved lines represent $-CH_2=O$ groups.)

20.90 Since both are the *cis* isomer, they are identical. One can be superimposed on the other after simple rotation.

20.92

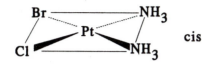 cis

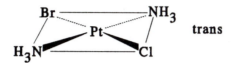 trans

20.94

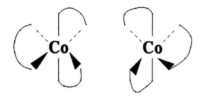

20.96 (a) $Cr(H_2O)_6^{3+}$ (b) $Cr(en)_3^{3+}$

20.98 $Cr(CN)_6^{3-}$

20.100 (a) The value of Δ increases down a group. Therefore, we choose: $[RuCl(NH_3)_5]^{3+}$
(b) The value of Δ increases with oxidation state of the metal. Therefore, we choose: $[Ru(NH_3)_6]^{3+}$

20.102 This is the one with the strongest field ligand, since Co^{2+} is a d^7 ion: CoA_6^{3+}

20.104 This is a weak field complex of Co^{2+}, and it should be a high–spin d^7 case. It cannot be diamagnetic; even if it were low spin, we would still have one unpaired electron.

Additional Exercises

20.110 (a)   The number of moles of chloride that have been precipitated is:

$$\# \text{ mol AgCl} = (0.538 \text{ g AgCl})\left(\frac{1 \text{ mol AgCl}}{143.32 \text{ g AgCl}}\right) = 3.75 \times 10^{-3} \text{ mol AgCl}$$

The number of moles of Cr that were originally present is:

$$\# \text{ mol Cr} = (0.500 \text{ g CrCl}_3 \cdot 6H_2O)\left(\frac{1 \text{ mol CrCl}_3 \cdot 6H_2O}{266.4 \text{ g CrCl}_3 \cdot 6H_2O}\right)\left(\frac{1 \text{ mol Cr}}{1 \text{ mol CrCl}_3 \cdot 6H_2O}\right)$$

$$= 1.88 \times 10^{-3} \text{ mol Cr}$$

The ratio of moles of Cl$^-$ per mole of Cr is therefore: 3.75/1.88 = 1.99.

This means that there were 2 mol of Cl$^-$ that were free to precipitate. The other mole of chloride ion must have been bound as a ligand to the Cr. In other words, the complex ion was $[Cr(Cl)(H_2O)_5]^{2+}$.

(b) $[Cr(Cl)(H_2O)_5]Cl_2\bullet H_2O$

(c)

(d)   There is only one isomer.

21.88   a) $AlP(s) + 3H_2O \rightarrow Al(OH)_3(s) + PH_3(g)$

         b) $Mg_2C(s) + 4H_2O(\ell) \rightarrow CH_4(g) + 2Mg(OH)_2(s)$

         c) $FeS(s) + 2HCl(aq) \rightarrow FeCl_2(aq) + H_2S(aq)$

         d) $MgSe(s) + H_2SO_4(aq) \rightarrow H_2Se(aq) + MgSO_4(aq)$

21.90   a) $2CO + O_2 \rightarrow 2CO_2$

         b) $2C_2H_6 + 7O_2 \rightarrow 4CO_2 + 6H_2O$

         c) $P_4O_6 + 2O_2 \rightarrow P_4O_{10}$

         d) $4NH_3 + 5O_2 \rightarrow 4NO + 6H_2O$

21.92 $\Delta G = 0 + 2(0) - 2(51.9 \text{ kJ/mol}) = -103.8 \text{ kJ}$

$$K_p = \frac{(P_{N_2})(P_{O_2})^2}{(P_{NO_2})}$$

$$\Delta G = -RT\ln K_p$$

$$K_p = \exp\left(-\frac{\Delta G}{RT}\right) = \exp\left(\frac{103.8 \times 10^3 \text{ J/mol}}{(8.314 \text{ J/molK})(298 \text{ K})}\right) = 1.57 \times 10^{18}$$

Because K_p is so large, we expect the equilibrium to lie to the right. Since we observe that NO_2 is stable, we must conclude that the reaction kinetics are too slow to achieve equilibrium.

21.94 $\Delta H = 9.7\text{kJ} - 2(33.8 \text{ kJ}) = -57.9 \text{ kJ}$

The reaction is exothermic so heat can be considered to be a product of the reaction. If we increase the temperature, effectively adding heat (a product), Le Chatelier's Principle states that the equilibrium will shift to the left and N_2O_4 will dissociate.

21.96   In each case, describe the structure without any hydrogen atoms, i.e., describe the anion structure.

        (a) IO_3^- is pyramidal

        (b) ClO_2^- is bent

        (c) IO_6^- is octahedral

        (d) ClO_4^- is tetrahedral

21.98   (a)   pyramidal

        (b)   t-shaped

        (c)   octahedral

        (d)   planar triangular

        (e)   square planar

Additional Exercises

21.100 $N_2 + H_2 \rightarrow 2NH_3$

          $4NH_3 + 5O_2 \rightarrow 4NO + 6H_2O$

          $2NO + O_2 \rightarrow 2NO_2$

          $3NO_2 + H_2O \rightarrow 2HNO_3 + NO$

21.102

Practice Exercises

P22.1 $\quad ^{226}_{88}\text{Ra} \rightarrow {}^{222}_{86}\text{Rn} + {}^{4}_{2}\text{He} + {}^{0}_{0}\gamma$

P22.2 $\quad ^{90}_{38}\text{Sr} \rightarrow {}^{90}_{39}\text{Y} + {}^{0}_{-1}\text{e}$

P22.3 $\quad$ We make use of the Inverse Square Law:

$$\frac{I_1}{I_2} = \frac{d_2^{\;2}}{d_1^{\;2}}$$

$$\frac{1.4 \text{ units}}{I_2} = \frac{(1.2 \text{ m})^2}{(10 \text{ m})^2}$$

$$I_2 = 100 \text{ units}$$

Review Exercises

22.43 $\quad$ Solve the Einstein equation for Δm:
$\Delta m = \Delta E/c^2$
$1 \text{ kJ} = 1.00 \text{ X } 10^3 \text{ J} = 1.00 \text{ X } 10^3 \text{ kg m}^2 \text{ s}^{-2}$
$\Delta m = 1.00 \text{ X } 10^3 \text{ kg m}^2 \text{ s}^{-2} \div (3.00 \text{ X } 10^8 \text{ m/s})^2 = 1.11 \text{ X } 10^{-14} \text{ kg} = 1.11 \text{ X } 10^{-11} \text{ g}$

22.45 $\quad$ The joule is equal to one kg m^2/s^2, and this is employed directly in the Einstein equation: $\Delta m = \Delta E/c^2$, where ΔE is the enthalpy of formation of liquid water, which is available in Table 5.2.

$H_2(g) + O_2(g) \rightarrow H_2O(\ell), \quad \Delta H = -285.9 \text{ kJ/mol}$
$\Delta m = (-285.9 \text{ X } 10^3 \text{ kg m}^2/\text{s}^2) \div (3.00 \text{ X } 10^8 \text{ m/s}^2) = -3.18 \text{ X } 10^{-12} \text{ kg}$
$-3.18 \text{ X } 10^{-12} \text{ kg} \times 1000 \text{ g/kg} \times 10^9 \text{ ng/g} = -3.18 \text{ ng}$
The negative value for the mass implies that mass is lost in the reaction.

22.47 $\quad$ The mass of the deuterium nucleus is the mass of the proton (1.00727252 u) plus that of a neutron (1.008665 u), or 2.015938 u. The difference between this calculated value and the observed value is equal to Δm:
$\Delta m = (2.015938 - 2.0135) = 2.4 \text{ X } 10^{-3} \text{ u}$
$\Delta E = \Delta mc^2 = (2.4 \text{ X } 10^{-3} \text{ u})(1.6606 \text{ X } 10^{-27} \text{ kg/u})(3.00 \text{ X } 10^8 \text{ m/s})^2$
$\Delta E = 3.6 \text{ X } 10^{-13} \text{ kg m}^2/\text{s}^2 = 3.6 \text{ X } 10^{-13} \text{ J}$

Since there are two neucleons per deuterium nucleus, we have:
$\Delta E = 3.6 \text{ X } 10^{-13} \text{ J/2 nucleons} = 1.8 \text{ X } 10^{-13} \text{ J per nucleon}$

22.49 $\quad$ (a) $^{211}_{83}\text{Bi}$ $\qquad$ (b) $^{177}_{72}\text{Hf}$ $\qquad$ (c) $^{216}_{84}\text{Po}$ $\qquad$ (d) $^{19}_{9}\text{F}$

22.51 $\quad$ (a) $\qquad ^{242}_{94}\text{Pu} \rightarrow {}^{4}_{2}\text{He} + {}^{238}_{92}\text{U}$

$\qquad$ (b) $\qquad ^{28}_{12}\text{Mg} \rightarrow {}^{0}_{-1}\text{e} + {}^{28}_{13}\text{Al}$

$\qquad$ (c) $\qquad ^{26}_{14}\text{Si} \rightarrow {}^{0}_{1}\text{e} + {}^{26}_{13}\text{Al}$

$\qquad$ (d) $\qquad ^{37}_{18}\text{Ar} + {}^{0}_{-1}\text{e} \rightarrow {}^{37}_{17}\text{Cl}$

22.53 $\quad$ (a) $^{261}_{102}\text{No}$ $\qquad$ (b) $^{211}_{82}\text{Pb}$ $\qquad$ (c) $^{141}_{61}\text{Pm}$ $\qquad$ (d) $^{179}_{74}\text{W}$

22.55 $^{87}_{36}Kr \rightarrow \, ^{86}_{36}Kr + \, ^{1}_{0}n$

22.57   The more likely process is positron emission, because this produces a product having a higher neutron–to–proton ratio: $^{38}_{19}K \; \rightarrow \; ^{0}_{1}e \; + \; ^{38}_{18}Ar$.

22.59   Six half life periods correspond to the fraction 1/64 of the initial material. That is, one sixty–fourth of the initial material is left after 6 half lives: 3.00 mg 1/64 = 0.0469 mg remaining.

22.61 $^{53}_{24}Cr^{*}$; $^{51}_{23}V \; + \; ^{2}_{1}H \; \rightarrow \; ^{53}_{24}Cr^{*} \; \rightarrow \; ^{1}_{1}p \; + \; ^{52}_{23}V$.

22.63 $^{80}_{35}Br$

22.65 $^{55}_{26}Fe$; $^{55}_{25}Mn \; + \; ^{1}_{1}p \; \rightarrow \; ^{1}_{0}n \; + \; ^{55}_{26}Fe$.

22.67 $^{70}_{30}Zn \; + \; ^{208}_{82}Pb \; \rightarrow \; ^{278}_{112}Uub \; \rightarrow \; ^{1}_{0}n + \, ^{277}_{112}Uub$

22.69

$$Radiation \, \alpha \, \frac{1}{d^2}$$

$$\frac{I_1}{I_2} = \frac{d_2^{\,2}}{d_1^{\,2}}$$

$$d_2 = d_1 \sqrt{\frac{I_1}{I_2}} = 2.0 m \sqrt{\frac{2.8}{0.28}} = 6.3 \, m$$

22.71   This calculation makes use of the Inverse Square Law:

$$\frac{I_1}{I_2} = \frac{d_2^{\,2}}{d_1^{\,2}}$$

$$\frac{8.4 \, rem}{0.50 \, rem} = \frac{d_2^{\,2}}{(1.60 \, m)^2}$$

$$d_2 = 6.6 \, m$$

22.73   The chemical product is $BaCl_2$. Recall that for a first order process k = 0.693/t1/2
So k = 0.693/30 yr = 2.30 x 10^{-2}/yr. Also,

$$\ln \frac{[A]_o}{[A]_t} = kt$$

$$[A]_t = [A]_o \, \exp(- kt)$$

$$= \exp[-(2.30 \times 10^{-2} \, /yr)(150 \, yr)]$$

$$= 3.17 \times 10^{-2}$$

so 3.1% of the original sample remains.

22.75   This calculation makes use of the first order rate equation, where knowing $[A]_t$, we need to calculate $[A]_o$:

$$\ln \frac{[A]_0}{[A]_t} = kt \, .$$

$k = 0.693/t_{1/2} = 0.693/8.07 \text{ d} = 8.59 \times 10^{-2} \text{ d}^{-1}$

$$\ln \frac{[A]_0}{\left(25.6 \times 10^{-5} \text{ Ci}/\text{g}\right)} = \left(8.59 \times 10^{-2} \text{ d}^{-1}\right)\left(28.0 \text{ d}\right)$$

Taking the antiln of both sides of the above equation gives:

$$\frac{[A]_0}{\left(25.6 \times 10^{-5} \text{ Ci}/\text{g}\right)} = e^{2.41} = 11.1$$

Solving for the value of $[A]_0$ gives: $[A]_0 = 2.84 \times 10^{-3}$ Ci/g

22.77   In order to solve this problem, it must be assumed that all of the argon–40 that is found in the rock must have come from the potassium–40, i.e., that the rock contains no other source of argon–40. If the above assumption is valid, then any argon–40 that is found in the rock represents an equivalent amount of potassium–40, since the stoichiometry is 1:1. Since equal amounts of potassium–40 and argon–40 have been found, this indicates that the amount of potassium–40 that remains is exactly half the amount that was present originally. In other words, the potassium–40 has undergone one half life of decay by the time of the analysis. The rock is thus seen to be 1.3×10^{9} years old.

22.79   Using equation 22.2 we may determine how long it has been since the tree died.

$$\frac{^{14}\text{C}}{^{12}\text{C}} = 1.2 \times 10^{-12} \, e^{-t/8270}$$

Taking the natural log we determine:

$$\ln\left(\frac{4.8 \times 10^{-14}}{1.2 \times 10^{-12}}\right) = -t/8270$$

$$t = -8270 \times \ln\left(\frac{4.8 \times 10^{-14}}{1.2 \times 10^{-12}}\right) = 2.7 \times 10^{4} \text{ yr}$$

The tree died 2.7×10^{4} years ago. This is when the volcanic eruption occurred.

22.81 $^{235}_{92}\text{U} + {}^{1}_{0}\text{n} \rightarrow {}^{94}_{38}\text{Sr} + {}^{140}_{54}\text{Xe} + 2\,{}^{1}_{0}\text{n}$.

Additional Exercises

22.92   Recall that for a first order process $k = 0.693/t_{1/2}$
So $k = 0.693/1.3 \times 10^{9}$ yr $= 5.30 \times 10^{-10}$/yr. Also,

$$\ln \frac{[A]_o}{[A]_t} = kt$$

$$t = \frac{1}{k} \ln \frac{[A]_o}{[A]_t} = \frac{1}{5.30 \times 10^{-10} \text{ /yr}} \ln \frac{3.22 \times 10^{-5} \text{ mol}}{2.07 \times 10^{-5} \text{ mol}} = 8.2 \times 10^{8} \text{ yrs}$$

22.94   Because this is an equilibrium process, there is a chance that either the forward or the reverse reactions will occur. As NO reacts with NO_2 we form ONO*NO. This compound can decompose and give either NO_2 and *NO or NO and *NO_2. After sufficient time, both *NO or NO will be present in the mixture.

22.97 This problem is similar to a dilution problem, i.e., $C_1V_1 = C_2V_2$. We will use cpm as the concentration unit. First, we need to account for the density difference.

$$\text{Methanol : concentration} = \left(\frac{580\,\text{cpm}}{\text{g}}\right)\left(\frac{0.792\,\text{g}}{\text{mL}}\right) = 460\,\frac{\text{cpm}}{\text{mL}}$$

$$\text{Coolant : concentration} = \left(\frac{29\,\text{cpm}}{\text{g}}\right)\left(\frac{0.884\,\text{g}}{\text{mL}}\right) = 26\,\frac{\text{cpm}}{\text{mL}}$$

Now we want the volume of the cooling system. Solve the following :

$$\frac{\left(\dfrac{459\,\text{cpm}}{\text{mL}}\right)(10.0\,\text{mL})}{\dfrac{26\,\text{cpm}}{\text{mL}}} = 180\,\text{mL}$$

Chapter Twenty-Three

Practice Exercises

P23.1 (a) 3–methylhexane
       (b) 4–ethyl–2,3–dimethylheptane
       (c) 5–ethyl–2,4,6–trimethyloctane

P23.2 (a)

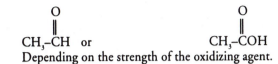

Depending on the strength of the oxidizing agent.

$$\underset{\underset{O}{\|}}{}$$

(b) CH₃CH₂CCH₂CH₃

Review Problems

23.82 (a)

```
         H     H
         |     |
 H —— C —— N —— H
         |
         H
```

(b)

```
               H
               |
 Br —— C —— Br
               |
               H
```

(c)

```
         Cl
         |
 H —— C —— Cl
         |
         Cl
```

(d)

```
         H
         |
 H —— N —— O —— H
```

(e)

```
 H —— C ≡ C —— H
```

(f)

```
         H     H
         |     |
 H —— N —— N —— H
```

23.84   (a) alkene   (d) carboxylic acid
        (b) alcohol   (e) amine
        (c) ester   (f) alcohol

        (a) alkyne   (d) ether
        (b) aldehyde   (e) amine
        (c) ketone   (f) amide

23.86   The saturated compounds are b, e, and f.

23.88   (a) amine   (b) amine   (c) amide   (d) amine, ketone

23.90   (a) These are identical, being oriented differently only.
        (b) These are identical, being drawn differently only.
        (c) These are unrelated, being alcohols with different numbers of carbon atoms.
        (d) These are isomers, since they have the same empirical formula, but different structures.
        (e) These are identical, being oriented differently only.
        (f) These are identical, being drawn differently only.
        (g) These are isomers, since they have the same empirical formula, but different structures.

23.92 (a) pentane
(b) 2–methylpentane
(c) 2,4–dimethylhexane

23.94 (a)   No isomers

(b)

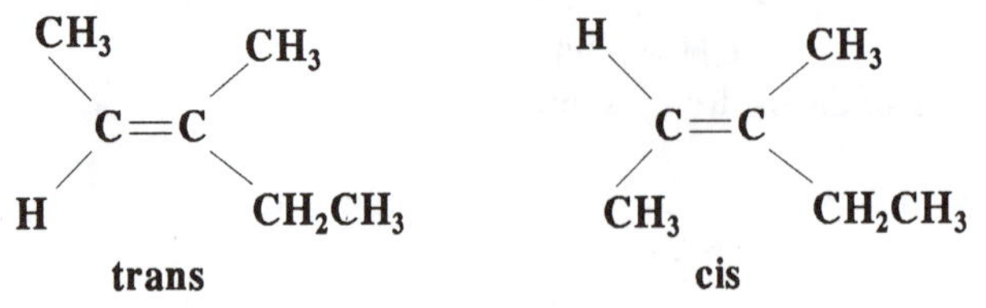

$$CH_3 \quad CH_3$$
$$C=C$$
$$H \qquad CH_2CH_3$$
trans

$$H \quad CH_3$$
$$C=C$$
$$CH_3 \qquad CH_2CH_3$$
cis

(c)

$$CH_2CH_3 \quad Cl$$
$$C=C$$
$$Cl \qquad CH_2CH_3$$
trans

$$Cl \quad Cl$$
$$C=C$$
$$CH_2CH_3 \quad CH_2CH_3$$
cis

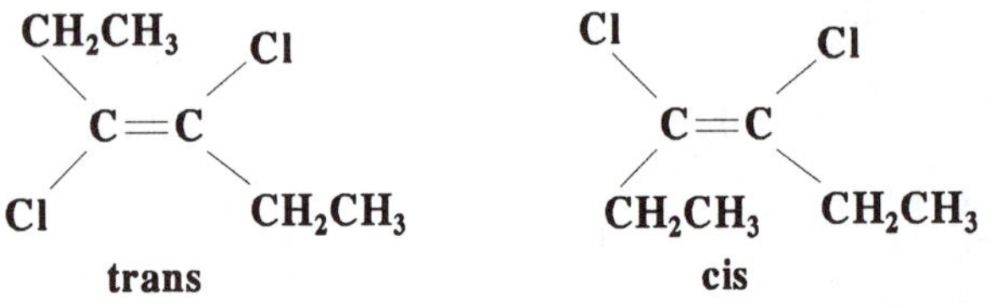

23.96 (a) CH_3CH_3
(b) $ClCH_2CH_2Cl$
(c) $BrCH_2CH_2Br$
(d) CH_3CH_2Cl
(e) CH_3CH_2Br
(f) CH_3CH_2OH

23.98 (a) $CH_3CH_2CH_2CH_3$

(b) $CH_3\ CH - CHCH_3$
              | |
              Cl   Cl

(c) $CH_3\ CH - CHCH_3$
              | |
              Br   Br

(d) $CH_3\ CH - CHCH_3$
              | |
              H   Cl

(e) $CH_3\ CH - CHCH_3$
              | |
              H   Br

(f) $CH_3\ CH - CHCH_3$
              | |
              H   OH

23.100 This sort of reaction would disrupt the delocalization of the benzene ring. The subsequent loss of resonance energy would not be favorable.

23.102 CH_3OH IUPAC name = methanol; common name = methyl alcohol
        CH_3CH_2OH IUPAC name = ethanol; common name = ethyl alcohol
        $CH_3CH_2CH_2OH$ IUPAC name = 1–propanol; common name = propyl alcohol
        $CH_3\ CHCH_3$ IUPAC name = 2–propanol; common name = isopropyl alcohol
           |
          OH

23.104   CH₃CH₂CH₂–O–CH₃   methyl ethyl ether
          CH₃CH₂–O–CH₂CH₃   diethyl ether
          (CH₃)₂CH–O–CH₃   methyl 2–propyl ether

$$CH_3CH_2CH_2\text{–}O\text{–}CH_3 \quad \text{methyl ethyl ether}$$
$$CH_3CH_2\text{–}O\text{–}CH_2CH_3 \quad \text{diethyl ether}$$
$$(CH_3)_2CH\text{–}O\text{–}CH_3 \quad \text{methyl 2–propyl ether}$$

23.106   (a)

        (b)

                $\text{C}_6\text{H}_5\text{—CH=CH}_2$

        (c)

                $\text{C}_6\text{H}_5\text{—CH=CH}_2$

23.108   (a)

        (cyclopentanone)

        (b)

                $\text{C}_6\text{H}_5\overset{\displaystyle O}{\overset{\|}{\text{—C-CH}_3}}$

        (c)

                $\text{C}_6\text{H}_5\text{-CH}_2\text{-}\overset{\displaystyle O}{\overset{\|}{\text{C-H}}}$

23.110   The elimination of water can result in a C=C double bond in two locations:
        CH₂=CHCH₂CH₃   CH₃CH=CHCH₃
        1–butene   2–butene

23.112   The aldehyde is more easily oxidized. The product is:
                   $\text{CH}_3\text{CH}_2\overset{\displaystyle O}{\overset{\|}{\text{COH}}}$

23.114 (a) $CH_3CH_2CO_2H$
(b) $CH_3CH_2CO_2H$ + CH_3OH
(c) Na^+ + $CH_3CH_2CH_2CO_2^-$ + H_2O

23.116 CH_3CO_2H + $CH_3CH_2NHCH_2CH_3$

23.118

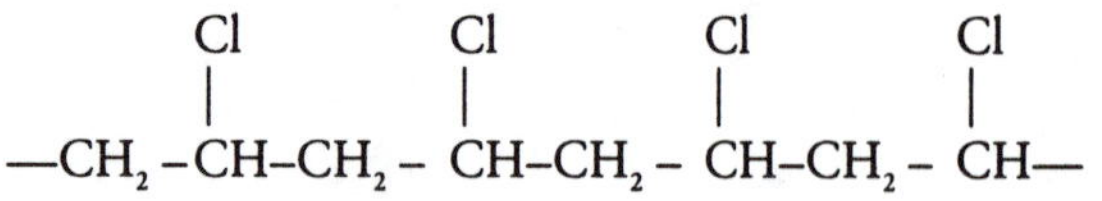

$$-CH_2-\underset{\underset{Cl}{|}}{CH}-CH_2-\underset{\underset{Cl}{|}}{CH}-CH_2-\underset{\underset{Cl}{|}}{CH}-CH_2-\underset{\underset{Cl}{|}}{CH}-$$

$$-(-CH_2CHCl-)-$$

23.120

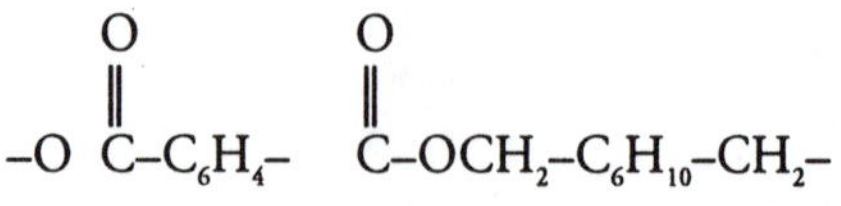

$$-O\overset{\overset{O}{\|}}{C}-C_6H_4-\quad\overset{\overset{O}{\|}}{C}-OCH_2-C_6H_{10}-CH_2-$$

23.122

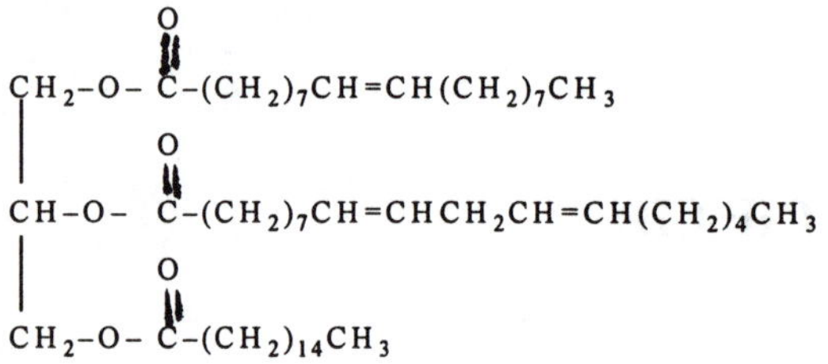

$$
\begin{array}{l}
CH_2-O-\overset{\overset{O}{\|}}{C}-(CH_2)_7CH=CH(CH_2)_7CH_3\\
|\\
CH-O-\overset{\overset{O}{\|}}{C}-(CH_2)_7CH=CHCH_2CH=CH(CH_2)_4CH_3\\
|\\
CH_2-O-\overset{\overset{O}{\|}}{C}-(CH_2)_{14}CH_3
\end{array}
$$

23.124

$$
\begin{array}{l}
CH_2-O-\overset{\overset{O}{\|}}{C}-(CH_2)_{16}CH_3\\
|\\
CH_2-O-\overset{\overset{O}{\|}}{C}-(CH_2)_{12}CH_3\\
|\\
CH_2-O-\overset{\overset{O}{\|}}{C}-(CH_2)_{16}CH_3
\end{array}
$$

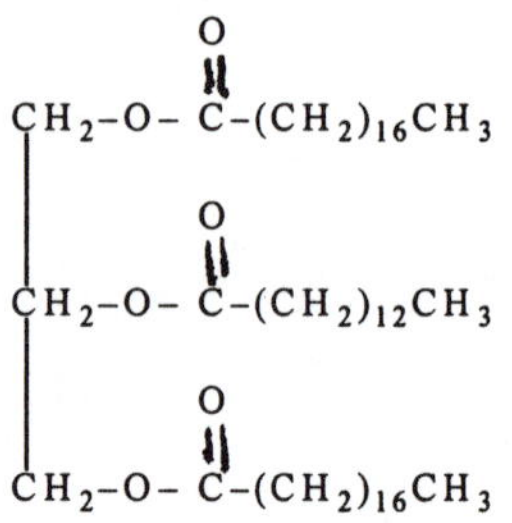

23.126 Hydrophobic sites are composed of fatty acid units. Hydrophilic sites are composed of charged units.

23.128

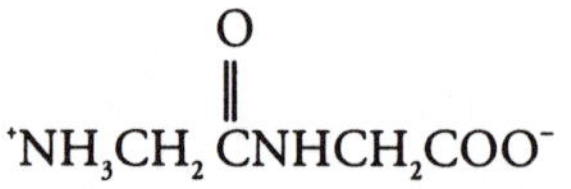

$$^+NH_3CH_2\overset{\overset{O}{\|}}{C}NHCH_2COO^-$$

23.130

$$^+NH_3\underset{\underset{CH_2C_6H_5}{|}}{CH}\overset{\overset{O}{\|}}{C}-NHCH_2\overset{\overset{O}{\|}}{C}OO^-\qquad ^+NH_3CH_2\overset{\overset{O}{\|}}{C}-NH-\underset{\underset{CH_2C_6H_5}{|}}{CH}\overset{\overset{O}{\|}}{C}OO^-$$

Additional Exercises

23.133 (a)   This must have been A, the primary alcohol, which is oxidized completely by an excess of dichromate to the acid. The secondary alcohol, B, would have been oxidized only to the ketone.

(b)

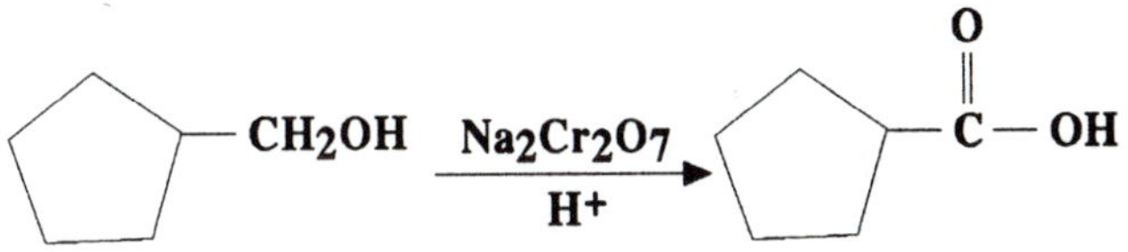

23.135 (a)

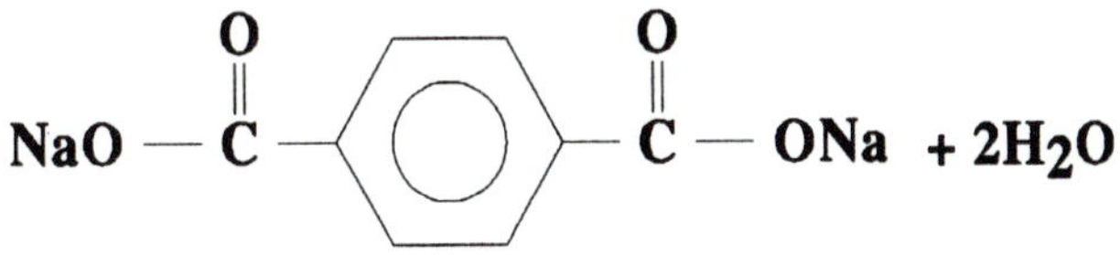

(b)

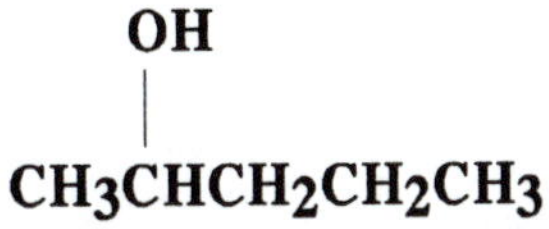

$$\text{CH}_3\overset{\overset{\displaystyle OH}{|}}{\text{CH}}\text{CH}_2\text{CH}_2\text{CH}_3$$

(c) CH$_3$NHCH$_2$CH$_2$CH$_3$ + H$_2$O
(d) CH$_3$CH$_2$OCH$_2$CH$_2$CO$_2$H + CH$_3$OH
(e)

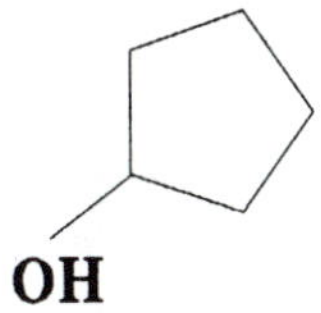

(f)

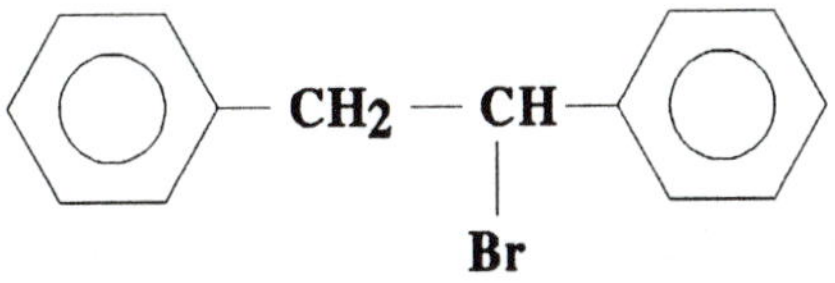

23.138 The original number of moles of hydroxide are:
0.1016 M 0.05000 L = 5.080 X 10^{-3} mol OH$^-$

The moles of hydroxide not neutralized are:
0.1182 M 0.02178 L = 0.002574 mol OH$^-$

Therefore, the moles of hydroxide that were neutralized by the acid were:
5.080 X 10^{-3} mol – 2.574 X 10^{-3} mol = 2.506 X 10^{-3} mol OH$^-$

This is also equal to the number of moles of the organic acid.

The formula mass is therefore:
0.2081 g/2.506 X 10^{-3} mol = 83.04 g/mol

This is equal to the molecular mass only if the unknown is a monoprotic acid.